SpringerBriefs in Cancer Research

For further volumes:
http://www.springer.com/series/10786

Seema Sethi

miRNAs and Target Genes in Breast Cancer Metastasis

 Springer

Seema Sethi
Department of Pathology
Wayne State University School of Medicine
    Karmanos Cancer Institute
Detroit, MI, USA

ISBN 978-3-319-08161-8   ISBN 978-3-319-08162-5 (eBook)
DOI 10.1007/978-3-319-08162-5
Springer Cham Heidelberg New York Dordrecht London

Library of Congress Control Number: 2014943913

Printed on acid-free paper

Springer is part of Springer Science+Business Media (www.springer.com)

To my parents, Sadanand and Tripta Bahri, who taught me the importance of sincere hard work in making a difference in this world. Losing my father to cancer and knowing the pain, anguish, and turmoil of this disease motivated me to work in this field. To my husband Anil and children Aisha, Prajit, and Sajiv who all make it worthwhile.

Contents

Chapter 1
Introduction: Role of miRNAs and Their Target Genes in Breast Cancer Metastasis

Seema Sethi, Shadan Ali, and Fazlul H. Sarkar

Keywords Breast cancer • miRNA • Brain metastasis • Bone metastasis

Breast cancer is the most common cancer among women in the United States and the second leading cause of cancer deaths among women of all ages [1]. In 2013, there have been approximately 232,340 new cases of invasive breast cancer and 39,620 breast cancer deaths among US women [2]. One out of 8 women in the United States will develop breast cancer in her lifetime [2].

Rapid advances in the fields of molecular biology and medicine have led to the development of novel therapeutic strategies for breast cancer. These have led to a significant improvement in the prognosis of this disease from the past few decades. Patients today have a wide range of therapeutic options including multimodality treatment protocols with surgery, chemotherapy, and molecular targeted therapies. Molecular-based therapies like trastuzumab, targeting against HER2/neu, have led to improved outcomes in these patients.

Although the prognosis has considerably improved for early stage cancers, unfortunately many patients die as a consequence of metastasis. It has been determined that approximately 25–40 % of patients develop metastatic disease which is generally incurable [3]. The metastasis could be at several body sites including the bone and brain. Once the metastasis develops, it heralds a rapid downhill course for these patients. Not only is the mortality increased but there is significant morbidity, impacting the quality of life of the patient.

Metastatic disease dramatically reduces the 5-year survival by 20 % when compared with patients with no metastasis [4]. Once breast cancer has

S. Sethi (✉) • F.H. Sarkar
Department of Pathology, Karmanos Cancer Institute, Wayne State University, Detroit, MI, USA
e-mail: drsethi7@gmail.com

S. Ali
Department of Oncology, Karmanos Cancer Institute, Wayne State University, Detroit, MI, USA

© Springer International Publishing Switzerland 2014
S. Sethi, *miRNAs and Target Genes in Breast Cancer Metastasis*,
SpringerBriefs in Cancer Research, DOI 10.1007/978-3-319-08162-5_1

metastasized, it becomes life threatening, the prognosis worsens, and patients require systemic treatment. Associated with reduction in the life span of patients are increasing side effects of therapies including chemo- and radiotherapeutic regimens to which the patient becomes unresponsive over time and the escalating healthcare costs, which become important for individual patients but become a major social and economic problem.

The breast cancer-related mortality and morbidity is primarily due to metastatic disease, especially metastatic disease into the brain and bone [5], which is a complex pathologic phenomenon occurring through a stepwise progression including invasion of surrounding stromal tissue, intravasation, evasion of programmed cell death (apoptosis), arrest in a vessel at a distant site, extravasation and subsequent establishment, and growth of the tumor at the site of metastasis (secondary growth in the metastatic milieu) [6–12]. The pathogenesis of these histological alterations in breast cancer is complex. The natural history of progression of breast cancers to cause brain and bone metastasis has several proposed mechanisms. However, the exact underlying molecular mechanisms are largely unknown.

Metastasis is a key hallmark of breast cancer and occurs when cancer cells access lymphatic and vascular systems and disseminate via lymph nodes and then via the venous and arterial vascular system to distant organs. Mechanistic insights into the pathologic development of metastasis at the molecular level could be helpful in understanding the key pathways implicated in this process. Understanding of the biological processes of metastasis would empower us with the clinical knowledge needed to identify and develop targets for implementing therapeutic and preventive strategies against future development of metastasis, with the eventual goal of improving patient survival and quality of life.

Another important aspect of recognizing metastasis is the time when the metastasis could occur. At present we cannot identify which patients will likely develop metastasis at which site and when. There is no fixed time period when this process begins. Metastatic relapse typically occurs many months to decades after surgery. Understanding of the processes that arise following tumor-cell dissemination including the phenomenon of dormancy would be helpful in early detection of metastasis in this disease, and identifying how tumor cells can be kept in a state of dormancy would provide strategies for management of this disease. Additional understanding is also needed in identifying the "pre-metastatic niches" in organs destined to develop metastases, which are proposed to generate metastases.

Further alterations at the molecular level have also been used for the molecular subclassification of breast cancer [13–17]. Certain tumor characteristics, e.g., mesenchymal/stromal gene signatures, have been related to some breast cancer subtypes (e.g., triple negative breast tumors), bone metastasis, and resistance to neoadjuvant therapies [18]. The pathologic development and progression of breast cancer seems to be a process-in-continuum which involves several molecular alterations at the genetic, epigenetic, and miRNA level, leading to a clonal evolution of malignant cells, and this process continues during metastatic progression. Breast oncogenesis, tumor progression, and development of metastasis involve deregulation of several processes.

Despite years of research and rapid advances in the fields of both molecular biology and medicine, the mystery of these processes has not yet been unraveled. Knowing the exact pathogenetic mechanisms implicated in breast cancer metastasis would lead to the identification of "actionable" molecular targets to ultimately improve the prognosis and survival of patients. In addition, identification of molecular alterations in patients with metastasis may also help determine prognosis and assist in risk stratification and would help design novel molecular targeted therapies to prevent and eliminate metastasis.

MicroRNAs (miRNAs) are recently described as small regulatory endogenous noncoding RNA molecules, approximately 18–25 nucleotides in length implicated in the posttranscriptional control of gene expression [19]. They are a major class of RNA molecules which regulate gene expression by targeting mRNAs to trigger either translational repression or degradation of mRNA. These tiny molecules are involved in developmental, physiologic phenomenon as well as pathologic processes including cancers [19]. In fact, miRNAs have emerged as critical regulators of cancer progression, invasion, and metastasis. This is mainly because a single miRNA can affect several downstream genes and signaling pathways with oncogenic or tumor suppressor actions depending on the target genes affected [19].

The miRNA expression levels have been found to be altered in several tumors [20–23]. Their expression levels may be either down- or upregulated in different cancers. A panel of altered miRNA expression levels—both the overexpressed miRNAs and the downregulated miRNAs—has the clinical potential to serve as a unique tumor-specific "signature." Such tumor-specific miRNA signatures can be used as biomarkers for screening and early diagnosis, prognosis, and risk stratification and as molecular targets for personalized medicine against tumors. Once the tumors are under treatment, these "signatures" can serve as biomarkers for tumor surveillance and recurrence.

Recent clinical trials have demonstrated the efficacy of personalized cancer therapy in improving the overall response rate and survival in cancer patients [24]. The miRNAs have the potential for clinical use as targets for personalized cancer therapy in human cancers. Targeting aberrantly expressed miRNAs using oligonucleotides and synthetic and natural agents holds a great promise as a novel targeted therapeutic approach to achieve the goal of personalized cancer therapy. In experimental models, synthesized oligonucleotides have been experimentally demonstrated to silence overexpressed miRNAs [25]. Synthetic and natural agents have also been demonstrated to regulate the expression of miRNAs. Recent studies suggest a dire need for evaluation of miRNAs in specific tumors and their metastasis for personalized cancer therapy [26].

The miRNAs are involved in the posttranscriptional regulation of several key physiologic and pathologic processes including cancers through their downstream signaling effects on several key genes [19]. The effects of miRNAs are orchestrated via a variety of mRNAs which degrade or inhibit further translation to proteins. Depending upon their target effect, miRNAs may play an oncogenic role or a tumor suppressive role. The classical examples of miRNAs exerting an oncogenic effect are miR-21, miR-17-92, miR-155, miR-221, and miR-222 which are overexpressed in cancers [19]. Let-7 family of miRNAs, miR-15, miR-16, miR-17-5p, miR-29, miR-34, miR-124a, miR-127, miR-143, miR-145, and miR-181 are examples of tumor suppressor miRNAs which are downregulated in cancers [19, 27].

The unique propensity of one miRNA to impact several downstream genes through its signaling pathways makes investigating the role of miRNAs in the patient context even more relevant [19]. Due to this multifold cascade effect, miRNAs have been proposed to be significant small endogenous molecules holding great promise in the clinical scenario for metastasis prevention. Recent studies have emphasized the potential role of miRNAs for targeted cancer therapeutics [28, 29]. Modulating the activity of miRNAs can provide opportunities for novel cancer interventions.

Several experiments are currently under way to exponentially understand the gene targets and signaling pathways orchestrated by the miRNAs [23]. These would assist in exploiting the complete spectrum of miRNA utility in the clinical realm including cancer accuracy and early diagnosis, risk stratification, and prognosis and as targets for anticancer therapies [30]. Targeting miRNAs could become a novel prognostic and therapeutic strategy to prevent the future development of metastasis [31]. Thus miRNAs could also serve as potential targets for antimetastatic therapy.

In breast cancer, several miRNAs have been implicated in the regulation of key carcinogenic events including cell cycle regulation and development of metastasis. These have led to a paradigm shift in the evaluation of the molecular regulators of cancer. Although present in physiologically normal states, these miRNAs are found to be altered in expression levels in cancers. Recent studies have demonstrated that miRNAs can be evaluated in a variety of clinical cancer specimens including fine needle aspirates of tumors increasing the utility of miRNAs in the clinical realm [21]. The expression of miRNAs in the primary tumor could be silenced using antagomirs (chemically modified anti-miRNA oligonucleotides) or treated with miRNA mimics for inducing its expression for the prevention and the development of metastasis. Therefore, development of miRNA-based prophylactic therapies could serve as precision and personalized medicine against future development of metastasis of breast and other cancers which will ultimately improve patients' quality of life, reduce healthcare costs, and improve overall survival.

Current experimental studies have demonstrated the efficacy of miRNAs in altering key pathways in several cancers [32]. The altered miRNA signatures can then regulate important genes in leading to the acquisition of key alterations in the cancer cells which can be targeted by different therapeutic agents [22]. The miRNAs have been recently described as novel molecules with implications and potential utility in the clinical arena including early diagnosis, prognosis, risk stratification, prevention of tumor progression, and treatment [30]. In breast cancers, the miRNA alterations have been shown to alter the physiologic cell structure, e.g., acquisition of the epithelial–mesenchymal phenotype, which enable them to acquire an invasive and metastatic capability [20]. Additionally, the altered miRNAs also have their clinical impact on the patients through regulation of cancer stem cells [8]. The effect of miRNAs on the cancer stem cells has far reaching consequences on tumor aggressiveness and metastatic potential [19].

Unique miRNA alterations identified in breast cancer metastasis would be helpful not only in early diagnosis but also in the determination of prognosis and in stratifying which miRNAs can be used as targets for neoadjuvant therapy. As we embark into the era of personalized and precision medicine, such efforts are becoming more important. Once we gain deep insights into the physiologic and pathologic processes involving breast cancer metastasis, we can address important treatment issues and prevention possibilities.

References

1. Punglia RS, Morrow M, Winer EP, Harris JR (2007) Local therapy and survival in breast cancer. N Engl J Med 356:2399–2405
2. DeSantis C, Ma J, Bryan L, Jemal A (2014) Breast cancer statistics, 2013. CA Cancer J Clin 64:52–62
3. Guarneri V, Conte P (2009) Metastatic breast cancer: therapeutic options according to molecular subtypes and prior adjuvant therapy. Oncologist 14:645–656
4. Chau NM, Ashcroft M (2004) Akt2: a role in breast cancer metastasis. Breast Cancer Res 6:55–57
5. Weil RJ, Palmieri DC, Bronder JL, Stark AM, Steeg PS (2005) Breast cancer metastasis to the central nervous system. Am J Pathol 167:913–920
6. Fernando RI, Litzinger M, Trono P, Hamilton DH, Schlom J, Palena C (2010) The T-box transcription factor Brachyury promotes epithelial-mesenchymal transition in human tumor cells. J Clin Invest 120:533–544
7. Hanahan D, Weinberg RA (2000) The hallmarks of cancer. Cell 100:57–70
8. Kong D, Banerjee S, Ahmad A, Li Y, Wang Z, Sethi S et al (2010) Epithelial to mesenchymal transition is mechanistically linked with stem cell signatures in prostate cancer cells. PLoS One 5:e12445
9. Nguyen DX, Massague J (2007) Genetic determinants of cancer metastasis. Nat Rev Genet 8:341–352
10. Sethi S, Macoska J, Chen W, Sarkar FH (2010) Molecular signature of epithelial-mesenchymal transition (EMT) in human prostate cancer bone metastasis. Am J Transl Res 3:90–99
11. Sethi S, Sarkar FH, Ahmed Q, Bandyopadhyay S, Nahleh ZA, Semaan A et al (2011) Molecular markers of epithelial-to-mesenchymal transition are associated with tumor aggressiveness in breast carcinoma. Transl Oncol 4:222–226
12. Vernon AE, Bakewell SJ, Chodosh LA (2007) Deciphering the molecular basis of breast cancer metastasis with mouse models. Rev Endocr Metab Disord 8:199–213
13. Baird RD, Caldas C (2013) Genetic heterogeneity in breast cancer: the road to personalized medicine? BMC Med 11:151
14. Hsieh SM, Look MP, Sieuwerts AM, Foekens JA, Hunter KW (2009) Distinct inherited metastasis susceptibility exists for different breast cancer subtypes: a prognosis study. Breast Cancer Res 11:R75
15. Hu G, Kang Y, Wang XF (2009) From breast to the brain: unraveling the puzzle of metastasis organotropism. J Mol Cell Biol 1:3–5
16. Nguyen DX, Bos PD, Massague J (2009) Metastasis: from dissemination to organ-specific colonization. Nat Rev Cancer 9:274–284
17. Reis-Filho JS, Pusztai L (2011) Gene expression profiling in breast cancer: classification, prognostication, and prediction. Lancet 378:1812–1823
18. Marchini C, Montani M, Konstantinidou G, Orru R, Mannucci S, Ramadori G et al (2010) Mesenchymal/stromal gene expression signature relates to basal-like breast cancers, identifies bone metastasis and predicts resistance to therapies. PLoS One 5:e14131

19. Sethi S, Sarkar FH (2011) Evolving concept of cancer stem cells: role of micro-RNAs and their implications in tumor aggressiveness. J Carcinogene Mutagene 17S1:005
20. Ahmad A, Aboukameel A, Kong D, Wang Z, Sethi S, Chen W et al (2011) Phosphoglucose isomerase/autocrine motility factor mediates epithelial-mesenchymal transition regulated by miR-200 in breast cancer cells. Cancer Res 71:3400–3409
21. Ali S, Saleh H, Sethi S, Sarkar FH, Philip PA (2012) MicroRNA profiling of diagnostic needle aspirates from patients with pancreatic cancer. Br J Cancer 107:1354–1360
22. Kong D, Heath E, Chen W, Cher ML, Powell I, Heilbrun L et al (2012) Loss of let-7 up-regulates EZH2 in prostate cancer consistent with the acquisition of cancer stem cell signatures that are attenuated by BR-DIM. PLoS One 7:e33729
23. Sarkar S, Dubaybo H, Ali S, Goncalves P, Kollepara SL, Sethi S et al (2013) Down-regulation of miR-221 inhibits proliferation of pancreatic cancer cells through up-regulation of PTEN, p27(kip1), p57(kip2), and PUMA. Am J Cancer Res 3:465–477
24. Tsimberidou AM, Iskander NG, Hong DS, Wheler JJ, Falchook GS, Fu S et al (2012) Personalized medicine in a phase I clinical trials program: the MD Anderson Cancer Center initiative. Clin Cancer Res 18:6373–6383
25. Krutzfeldt J, Rajewsky N, Braich R, Rajeev KG, Tuschl T, Manoharan M et al (2005) Silencing of microRNAs in vivo with 'antagomirs'. Nature 438:685–689
26. Bao B, Azmi AS, Ali S, Ahmad A, Li Y, Banerjee S et al (2012) The biological kinship of hypoxia with CSC and EMT and their relationship with deregulated expression of miRNAs and tumor aggressiveness. Biochim Biophys Acta 1826:272–296
27. Sethi S, Kong D, Land S, Dyson G, Sakr WA, Sarkar FH (2013) Comprehensive molecular oncogenomic profiling and miRNA analysis of prostate cancer. Am J Transl Res 5:200–211
28. Bao B, Ali S, Ahmad A, Azmi AS, Li Y, Banerjee S et al (2012) Hypoxia-induced aggressiveness of pancreatic cancer cells is due to increased expression of VEGF, IL-6 and miR-21, which can be attenuated by CDF treatment. PLoS One 7:e50165
29. Sethi S, Li Y, Sarkar FH (2013) Regulating miRNA by natural agents as a new strategy for cancer treatment. Curr Drug Targets 14:1167–1174
30. Sethi S, Ali S, Kong D, Philip PA, Sarkar FH (2013) Clinical implication of microRNAs in molecular pathology. Clin Lab Med 33:773–786
31. Sethi S, Ali S, Philip PA, Sarkar FH (2013) Clinical advances in molecular biomarkers for cancer diagnosis and therapy. Int J Mol Sci 14:14771–14784
32. Qazi AM, Gruzdyn O, Semaan A, Seward S, Chamala S, Dhulipala V et al (2012) Restoration of E-cadherin expression in pancreatic ductal adenocarcinoma treated with microRNA-101. Surgery 152:704–711

Chapter 2
Molecular Pathogenesis of Breast Cancer and the Role of MicroRNAs

Shadan Ali, Seema Sethi, Azfur S. Ali, Philip A. Philip, and Fazlul H. Sarkar

Abstract Breast cancer is a leading cause of cancer-associated death in women worldwide. The therapy usually involves mastectomy or lumpectomy, followed by chemotherapy and/or radiation therapy in addition to hormonal therapy when indicated. While the area of research on the identification and potential use of microRNAs (miRNAs) as either a diagnostic, prognostic, or predictive biomarker is still in its early stages, there is increasing evidence that miRNAs are involved in tumor progression, chemoresistance, and survival. The miRNAs have enormous prospective in clinical research since they are detected in the serum, plasma, fresh tissues, and formalin-fixed paraffin-embedded tissue samples. Hence, it may be possible to develop novel therapeutic regimens of specific miRNAs as targets to prevent or treat breast cancer (BC). The miRNA expression profiling is now used extensively by many investigators to demonstrate specific miRNA signatures in both the body fluids and in the tumor tissue, indicating that miRNAs may likely be useful as diagnostic and prognostic tools in all cancers including BC. Numerous investigators, including our laboratory, have used strategies to deregulate miRNAs with either anti- and pre-miRNA molecular drugs or even natural compounds to prevent or control tumor progression, which will be discussed in this chapter. Moreover, the role of several natural and synthetic compounds as anticancer agents will also be discussed in this chapter. Finally, the role of several miRNAs as targets will be discussed especially because miRNA-based therapies are currently being exploited for cancer therapy.

S. Ali • P.A. Philip
Department of Oncology, Karmanos Cancer Institute, Wayne State University School of Medicine, Detroit, MI, USA

S. Sethi • A.S. Ali
Department of Pathology, Karmanos Cancer Institute, Wayne State University School of Medicine, 740 Hudson Webber Cancer Research Center, 4100 John R Street, Detroit, MI, USA

F.H. Sarkar (✉)
Department of Oncology, Karmanos Cancer Institute, Wayne State University School of Medicine, Detroit, MI, USA

Department of Pathology, Karmanos Cancer Institute, Wayne State University School of Medicine, 740 Hudson Webber Cancer Research Center, 4100 John R Street, Detroit, MI, USA
e-mail: fsarkar@med.wayne.edu

© Springer International Publishing Switzerland 2014
S. Sethi, *miRNAs and Target Genes in Breast Cancer Metastasis*,
SpringerBriefs in Cancer Research, DOI 10.1007/978-3-319-08162-5_2

Keywords Breast cancer • Serum • FFPE • miR-21 • miR-125 • miR-34a • miR-200 • miR-155 • miR-10b • Plumbagin • Garcinol • Curcumin

Introduction

Breast adenocarcinoma (breast cancer, BC) is the most commonly diagnosed cancer among women in the United States with an estimated 232,340 new cases in 2013. Sadly, one in eight is diagnosed with breast cancer in their lifetime, yet it is still a frequent cause of death in women [1]. Although research together with progress in targeted therapies led to increased patient survival, there is still an increased demand for the development of new diagnostic biomarkers and groundbreaking therapeutic strategies for improving the overall survival in all patients diagnosed with BC. Recent research suggests the involvement of deregulated genes at the levels of DNA, protein, RNA, and microRNAs (miRNAs), and as such miRNAs are becoming important players in the development, differentiation, and regulation of gene expression in cancer biology [2, 3]. The miRNAs are small noncoding endogenous single-stranded class of regulatory RNAs that posttranscriptionally inhibit gene expression through targeting specific messenger RNAs (mRNAs) [4]. Due to the smaller size, these miRNAs remain stable in body specimens including plasma, serum, and both fresh-frozen and formalin-fixed paraffin-embedded (FFPE) tissue samples and can serve as an excellent source of early detection of various cancers, including BC [5, 6]. Emerging evidence suggests that a single miRNA may target several mRNAs and vice versa and substantial basic-science research on these miRNAs has led to the development of new methodologies for cancer diagnosis, and this is now progressing to clinical research arenas. Comparison of cancerous against normal human plasma or tissue samples by miRNA microarray showed deregulation of several miRNAs in many cancers such as lung, pancreas, prostate, and BC [5–10]. Among the many deregulated miRNAs, several of them are common in most cancers and few are cancer specific [11]. Research on antisense oligonucleotides (ASO) is also on the rise; use of this technology in vivo has been successfully demonstrated by delivering ASO with chemokines to inactivate a subset of immune cells [12] and in many other conditions. In this book chapter, we review the current and developing methodologies of extracting miRNA from body fluids and fresh-frozen and FFPE tissue samples and will discuss few miRNAs and their use in the diagnosis and treatment of BC.

Detection of microRNAs in Clinical Specimens

The challenge that clinicians face is in the early detection of cancer to reduce mortality rates. Significant research in the fields of molecular biology has led to the development of newly identified miRNAs which has implications in the prevention of disease progression and therapeutic targets for designing molecular therapies for the management of cancer patients. Many investigators including our own laboratory have shown miRNAs as suitable biomarkers for early detection of cancers since they are stable and not degraded in plasma, serum or fresh-frozen and fine-needle aspirates of FFPE tissue samples [5, 6]. The miRNA expression profiling has been helpful in differentiating normal patients from cancer patients due to differential expression of various miRNAs. Several miRNAs such as miR-34a, miR-155, and miR-10b have been shown to be deregulated in the serum of BC patients compared to healthy controls [13]. Another study showed differences in circulating miRNA level between Caucasian and African-American patients [14]. In addition, specific miRNAs may also predict drug response that is essential for developing precise molecular targeted therapy for each individual. Moreover, miRNAs can be up- or downregulated in cancers due to the genes' downstream signaling effect, hence controlling the expression of cancers [15]. The miRNAs that are elevated in cancer are oncogenic, and likewise, the miRNAs that have reduced expression function as tumor suppressor [16–19].

Methodology and Clinical Associations

Although new technologies have advanced cancer research from bench to bedside, it remains a common medical problem worldwide in all cancers leading to significant mortality, morbidity, and rising healthcare costs. This emphasizes the urgent need for novel molecular technologies both in the laboratory and in the clinic to identify high-risk patients. Small molecules, such as miRNAs, have enormous potential in the clinic since a single miRNA can target multiple genes, indicating that modulating one miRNA will have effects on multiple genes which is opening new doors for innovative therapies especially to target heterogeneous populations of cancer cells within a tumor mass. The discovery of noninvasive biomarkers for most cancer types has been a valuable tool to differentiate tumors from normal tissue with minimal discomfort and risk to the cancer patients. Their most important benefit is the ease of access and possibility of repeated testing in a noninvasive manner. Emerging evidence suggests that the expression level of miRNAs varies in various cancers and also changes as the disease progresses. Some miRNAs are tumor specific both in human and in mouse models [20], while others are common in many tumor types [21, 22].

Isolation of microRNAs, Reverse Transcription, and Polymerase Chain Reaction (RT-PCR) from Plasma Samples

A number of independent researchers studied the potential role of miRNAs in the plasma as early detection biomarkers of cancers including BC [5, 7, 11, 13]. The miRNAs have been found to be stable in many body fluids including plasma and serum which is believed to be important for their potential as disease biomarkers. The assay requires a very small amount of RNA as low as 10 ng which can be easily isolated from either plasma or serum for real-time PCR quantitative analysis. The detailed methodology of isolation of miRNAs from plasma has been described earlier [15]. Here we will highlight few important points and techniques. The total RNA including miRNAs is isolated from as low as 250 µl of plasma sample using QIAGEN kit (QIAGEN, Valencia, CA) and eluted in 25 µl of water. Although RNA obtained from plasma is low, and it cannot be quantified using single-drop NanoDrop technology, similar volume can be used together with housekeeping miRNAs as controls. Nonetheless, the reverse-transcription (RT) reaction can be carried out using the template mature miRNA by using Exiqon-Universal cDNA synthesis kit available from Exiqon. The RT reaction contains 4 µl of 5X RT buffer, 2 µl of enzyme, 4 µl of either plasma miRNA or 250 nM of standard miRNA, and 10 µl of water incubated for 1 h at 42 °C and 5 min at 95 °C.

The cDNA obtained from above is subjected to real-time polymerase chain reaction (PCR) using multiple housekeeping genes for data normalization. The analysis is performed by the standard Ct method for quantification using StepOnePlus Real-Time PCR (Applied Biosystems, Foster City, CA) which also serves as control for variability in sample loading. The miRNA standard cDNA and the plasma cDNA is diluted in water, and the reaction is set up using SYBR Green (Applied Biosystems) and PCR primer mix as described earlier [15].

MicroRNA Methodology Utilizing Archived Formalin-Fixed Paraffin-Embedded Tissues

The total RNA containing miRNA is isolated from FFPE tissue using RNeasy Kit (QIAGEN) following manufacturer's protocol using four 10-µm thick and approximately 0.5–1 cm in diameter tissue curls as described previously [15]. The total RNA containing miRNA is then eluted with RNase-free water and measured and quantified using NanoDrop 2000 (Thermo Scientific, Pittsburgh, PA). The RT reaction is performed with SYBR Green miRNA-based assay using Exiqon-Universal cDNA synthesis kit (Exiqon, Woburn, MA) using 10 ng of total RNA. PCR reactions are performed in triplicate using StepOnePlus Real-Time PCR (Applied Biosystems), and expression levels of miRNAs are analyzed using Ct method.

Developments Made in Differentiating Normal and Disease State Using miRNA Profiling

We have described RNA extraction methods both from plasma and FFPE tissues previously [15], and few additional points are also discussed in this chapter. The ultimate challenge or concern to miRNA extraction from plasma is the low level of circulating miRNAs which are below the detection limit of spectrophotometry. Compared to microarray profiling, qRT-PCR have shown superiority in sensitivity [23], and hence these methodologies are often used to validate abnormal expression of miRNAs. To avoid unfair measurement of miRNAs, endogenous genes are used as internal controls for data normalization [6]. Initial research on human breast tumors identified the variation in gene expression using RNA from 42 patients with complimentary DNA microarrays that showed great variation in gene expression, yet it also showed specific gene expression relating to tumor types [24]. In addition, gene expression profiling of hereditary BC patients discovered exclusive expression patterns that were dependent on the *BRCA1* and *BRCA2* mutation status [25]. Although gene expression profiling was the standard for defining molecular subtypes, immunohistochemical analysis was also used for expression of hormone receptors and lack of *HER2/neu* overexpression and luminal cytokeratin [26]. Several developing technologies are looking outside of gene expression profiling such as toward the level of protein expression or gene methylation to understand the differences associated between normal tissue and cancer [27]. Finally, the probability of using miRNA expression profiling in clinical samples by microarray or by quantitative RT-PCR emerged in several studies as diagnostic and prognostic marker [27, 28]. The miRNA research has made a substantial impact on efficient profiling of deregulated miRNAs in plasma, serum, FFPE, and many other sample types because of their stability, serving as a potentially reliable biomarker [5–7, 11, 13, 14]. Expression analysis of miRNAs not only determines several miRNAs but also fully discriminates between normal healthy and diseased state. There are a number of miRNAs that are substantially upregulated in one type of cancer; for instance, miR-205 is overexpressed in the lung, pancreas, and bladder cancer [13, 29–31] and found to be decreased in breast, prostate, and esophageal cancer [32–34], suggesting that some miRNAs act as both oncogenic and tumor suppressor depending on the tumor type and expression pattern. Therefore, the field of miRNA research is highly complex and requires critical insights with respect to the function of a specific miRNA in a specific context.

Altered miRNA Expression in Breast Cancer

Despite substantial improvements in the field of cancer biology, the progress of validated biomarkers for BC has remained an overwhelming task. Numerous studies have been published identifying deregulated miRNAs in BC using microarray profiling and subsequent validation of selected miRNAs by real-time PCR. The results of miRNA profiling suggest that many miRNAs are altered in all types of cancer and may provide a useful biomarker for detection of cancer. Overexpression of miRNAs in human BC is often a result of molecular genomic abnormalities, called OncomiRs. Inhibition of these OncomiRs can inhibit cell proliferation as well as tumor growth. Patients with triple negative BC (TNBC) have poor prognosis due to aggressive proliferation, migration, and invasion. One study with MDA-MB-231 parental cells and MDA-MB-231 cells stably expressing miR-221-ZIP or scramble-ZIP showed knockdown of miR-221 which inhibited tumor growth by altering the expression of E-cadherin, snail, and slug both in vivo and in TNBC cell lines in vitro [35]. In the subsequent section, we will discuss few miRNAs that are OncomiRs and tumor suppressors in BC.

miR-21

One well-described OncomiR globally found in many tumors including BC is miR-21. The important target of miR-21 is phosphatase and tensin homolog (PTEN) and PDCD4 [36]. Loss of PTEN has been found to be indirectly associated with miR-21 expression in the breast, pancreas, and colon cancer [36–39]. Iorio et al. reported aberrant expression of 29 miRNAs in breast cancer tissues using microarray analysis compared to normal tissues [7]. The miR-21 was also recognized as overexpressed miRNA in a large-scale miRNome analysis on 540 samples that included the lung, breast, stomach, prostate, colon, and pancreatic tumors [40]. By TaqMan real-time PCR methodology, miR-21 was proven to be upregulated in breast tumors compared to the normal breast tissue samples among 157 miRNAs analyzed [41]. A recent study demonstrated overexpression of miR-21 in FFPE tissue samples of atypical ductal hyperplasia, ductal carcinoma in situ, and invasive ductal carcinoma compared to normal tissue samples, suggesting its oncogenic role in all types of cancer including BC [42].

miR-155

The miR-155 has been shown to be upregulated in different tumor types, including BC [7, 43–45]. It directly inhibits RhoA expression, a gene that regulates cell adhesion, motility, and polarity [7]. It is also linked with cancer invasiveness in human BC [44]. Inhibition of miR-155 induced apoptosis and improved chemosensitivity in BC cell lines by targeting FOXO3a [43]. Inhibition of miR-155 with antisense oligonucleotide (ASO-miR-155) in MDA-MB-157 breast cancer cell line inhibited cell viability, induced apoptosis, and most importantly inhibited tumor growth in mouse model which was in part mediated via capase-3 upregulation [45].

miR-10b

The miR-10b is greatly expressed in hepatocellular, glioblastoma, pancreatic, and breast tumors [29, 46–48]. Chan et al. studied the expression of circulating miRNAs from Asian Chinese to compare miRNA expression from serum samples obtained from BC patients and healthy individuals using microarrays or locked nucleic acid real-time PCR panels. Among the significantly expressed miRNAs, miR-10b was significantly upregulated in serum of BC patients compared to serum obtained from healthy controls, suggesting the noninvasive diagnostic strategy could be a promising tool for clinical studies, although further validation in different subtypes of breast cancer is warranted [49]. Overexpression of miR-10b in metastatic BC cells regulated cell migration and invasion through the transcription factor Twist which, in turn, inhibited homeobox D10, resulting in increased expression of RHOC [47]. Higher expression of miR-10b in primary breast carcinomas was associated with clinical progression [47]. In addition, miR-10b overexpression was also observed in metastasis-positive patients compared to metastasis-free patients in hepatocellular carcinoma [48].

miR-34a

Overexpression of miR-34a decreases Akt signaling pathway and increases estrogen receptor-alpha (ERα)-phosphorylation status [50]. The expression of miR-34a is typically decreased in cancer causing activated signaling such as through Akt pathway. A recent article by Guo et al. stated upregulation of miR-34a with curcumin and its combination with another natural compound, emodin, led to the downregulation of Bcl-2 and Bmi-1 in breast cancer cells, indicating the involvement of both apoptosis regulator and self-renewal of adult stem cells [51]. Sensitization of MCF-7 cells to Adriamycin was also observed with ectopic overexpression of miR-34a, suggesting that deregulation of miR-34a plays a key role in acquired Adriamycin resistance of BC, to some extent by targeting Notch-1, another target of miR-34a [52]. Some miRNAs are differentially expressed in the blood of breast and colorectal cancer patients compared to controls. The analysis of the relative quantification of the miRNAs showed significantly reduced levels of expression of miR-34a both in breast and in colorectal cancer patients compared to controls, suggesting that miR-34a is not tissue specific and may be used in the future as a circulating biomarker for multiple cancers [53].

miR-125a,b

HDAC inhibitor entinostat inhibited erbB2/erbB3 protein translation through upregulation of miR-125a, miR-125b, and miR-205 via targeting erbB2 and/or erbB3 in BC cells [54]. Similarly, one investigator report using SKBR3 cells, a breast cancer cell line, as a model for ERBB2/ERBB3 dependence, showed through infection of cells with retroviral constructs expressing miR-125a or 125b which resulted in the inhibition of ERBB2/ERBB3, suggesting the possibility of using miRNAs as a beneficial strategy for therapeutic target [55]. Another report demonstrated that the expression of miR-125a was related with the expression of stress-induced RNA binding protein HuR, which is high in many cancers including BC. Restoration of miR-125a expression reduced HuR protein level and repressed cell growth in breast cancer cells, suggesting that miR-125a may play a role as a tumor suppressor in BC [56].

miR-200 Family

Several investigators have demonstrated the key role of miR-200 family in regulating epithelial-to-mesenchymal transition (EMT) via inhibition of the E-cadherin transcriptional repressors ZEB1/ZEB2 [57–61]. A recent study also demonstrated in a mouse model of BC metastasis that ectopic expression of the miR-200b/200c/429 limits tumor-cell invasion and metastasis [62]. Furthermore, moesin was found to be directly targeted by miR-200b, and thus restoration of miR-200b expression in cells alleviated metastatic suppression, suggesting the existence of a moesin-dependent pathway which is different from the ZEB1/ZEB2 pathway [62]. Micro-environmental signals including TGFβ can direct tumor metastasis by varying miR-200 expression [58]. The role of miR-200 as tumor suppressor was also studied by Manavalan et al. in BC cell line model of advancing endocrine/tamoxifen resistance. The study showed that overexpression of miR-200b or 200c in endocrine therapy (tamoxifen)-resistant cells changed morphology to epithelial appearance, inhibited cell growth and migration, and were sensitized to tamoxifen [63]. Tamoxifen effect was also observed in endometrial cancer cells, which was in part through upregulation of snail and downregulation of E-cadherin and miR-200 expression, contributing to tamoxifen-induced EMT through c-Myc [64]. Lim et al. in a recent study suggested that cancer stemlike cells show loss of miR-200 expression, and restoration of its expression decreased stemlike characteristics and further stimulated epithelial phenotype in BC cells [65]. Studies in other cancers also showed upregulation of E-cadherin caused by ectopic expression of miR-200 family, suggesting miR-200 as a marker of the epithelial phenotype [61]. Many investigators reported anticancer properties of natural and synthetic compounds, such as reduction in proliferation [37, 60, 66] and cancer cell-specific induction of apoptosis in BC cells mediated through deregulation of several signaling pathways

[66, 67]. In this chapter we will discuss few natural and synthetic anticancer compounds and their attributes as modulators of miRNAs.

Natural and Synthetic Anticancer Compounds

Garcinol

Anticancer properties of edible fruit *Garcinia indica*-derived garcinol are opening new doors for chemotherapy or chemoprevention of many cancers, including BC [67–71]. It was demonstrated earlier by our group that garcinol-induced apoptosis in the breast, prostate, and pancreatic cancer cells is mediated through the downregulation of NF-κB signaling pathway [67, 68]. Another recent report by our group showed the important role of garcinol in the reversal of EMT, as observed in aggressive triple negative MDA-MB-231 and BT-549 breast cancer cells which was mediated through the upregulation of epithelial marker E-cadherin and the expression of miR-200 and let-7 family miRNAs [69]. Another study investigated the effect of garcinol on a human hepatocellular cancer cell line Hep3B that lacks functional p53. Garcinol activated the mitochondrial apoptotic pathways along with the ER stress modulator GADD153, indicating a potential therapeutic role of garcinol in p53-independent apoptosis in cancer [71]. In addition, garcinol inhibited cell proliferation in nicotine-induced human BC, MDA-MB-231 cells, through the downregulation of α9-nAChR and cyclin D3 expression, suggesting that cyclin D3 would be a suitable molecular target for assessing the activity of chemotherapeutics and/or garcinol could be a powerful chemopreventive agent in BC patients in the clinical setting [70].

Plumbagin

Many investigators reported potent anticancer activity of a plant metabolite plumbagin, a naturally occurring naphthaquinone (5-hydroxy-2-methyl-1, 4-naphthoquinone) [66, 72–74]. Plumbagin has been shown to inhibit cancer cell migration and invasion and suppressed the expression of osteoclast-activating factors [66]. It also inhibited breast tumor-bone metastasis and osteolysis by controlling the tumor-bone microenvironment in a mouse model, suggesting that plumbagin may serve as an innovative agent for the treatment of tumor-bone metastasis [66]. Plumbagin can induce estrogen-dependent cell signaling and apoptosis in BRCA1-blocked ovarian cancer cells. It was observed to be the most effective anticancer agent when compared to other structurally related compounds and indicated to enhance numerous pathways of apoptosis and cell cycle arrest in BRCA1-blocked cells compared to unblocked cells [72]. Another investigator observed the potential role of plumbagin toward the expression of CXCR4 and its function in various tumor cells. It was observed that plumbagin downregulated CXCR4 expression in BC cells regardless of their HER2 status and was not cell type specific. For example, inhibition with plumbagin also occurred in the gastric, lung, renal, oral, and hepatocellular cancer cell lines [74]; however, no specific miRNA has been found to be associated with the biological activity of plumbagin.

Curcumin

Curcumin [1,7-bis-(4-hydroxy-3-methoxyphenyl)-1,6-heptadiene-3,5-dione] is a polyphenolic compound found in the spice turmeric and is considered as a pleio-tropic molecule which interacts with a variety of molecular targets and has antitumor, anti-inflammatory, and various other biological activities [75]. Curcumin inhibited the growth of BC cell lines, and increased the percentage of cells in sub-G0 phase, representing the apoptotic cell population, which was further con-firmed by PARP-1 cleavage [75]. In vivo antitumor activity of both early and in an advanced stage of mammary carcinogenesis induced tumor-free survival and a reduction in tumor multiplicity with the administration of safe curcumin [75]. Curcumin was observed to stabilize p27 levels in BC with concomitant decrease in Skp2, Her2, Cyclin E, and CDK kinases expression in MDA-MB-231/Her2 cells, suggesting the potential role of curcumin as a chemopreventive agent in BC [76]. Curcumin is not only considered as chemopreventive/chemother-apeutic drug but is also associated with obesity-related cancers. It modifies several molecular targets by reversing insulin resistance to prevent obesity-related cancers [77]. Demethoxycurcumin, an active compound of curcuminoids originated in turmeric powder, showed reduced levels of ECM degradation-associated proteins such as matrix metalloproteinase-9 (MMP-9), membrane type-1 matrix metalloproteinase (MT1-MMP), urokinase plasminogen activator (uPA), and uPA receptor (uPAR) curcumin-treated in MDA-MB-231 cells [78].

Although in vitro and some limited preclinical model study showed promising results with curcumin, human clinical trial has been disappointing which was partly attributed to the target tissue bioavailability of curcumin. The low bioavailability of curcumin prompted the synthesis of many analogs of curcumin. One such example is CDF, a difluorinated synthetic analog of curcumin with greater bioavailability [38]. Although the effect of CDF has not been demonstrated in BC yet, it has been proven to be more effective than curcumin in both pancreatic and colon cancer [37, 38, 79]. In order to ensure that the efficiency of CDF was similar to that of curcumin, tests were conducted in pancreatic cancer cell lines comparing the two in terms of their ability to inhibit cell growth both in vitro and in vivo [37]. CDF also reduced the presence of cancer stem cell markers in chemoresistant colon cancer cells compared to curcumin [79]. The above two reports confirmed that CDF was more effective than curcumin, and the biological activity was in part mediated through deregulation of miRNAs [38].

Table 2.1 The list of up- and downregulated microRNAs and their Targets

miRNAs	Up- or downregulated	Target genes	References
miR-221	Upregulated	E-cadherin, snail, slug	[35]
miR-21	Upregulated	PTEN, PDCD4	[36–40]
miR-155	Upregulated	RhoA, FOXO3a, Caspase-3	[7, 43, 45]
miR-10b	Upregulated	Twist, RHOC	[47]
miR-34a	Downregulated	Akt, Bcl2, Bmi-1, Notch1	[50–52]
miR-125a,b	Downregulated	erbB2, erbB3, HUR	[54–56]
miR-200 family	Downregulated	E-cadherin, ZEB1, ZEB2, c-Myc	[57–61, 64]

Conclusion

For the past decade or more, numerous studies have shown an intricate relationship between the expression of miRNA and human malignancies including BC. The miRNAs are important regulators of numerous biological processes and are implicated in the pathogenesis of not only cancers but also other human diseases. Recent studies have shown the stability of miRNAs in serum, plasma, and in both fresh and FFPE tissue samples. Microarray expression profiling of miRNAs provides a high-throughput molecular resource to classify countless number of diseases including BC associated with deregulated expression of miRNAs, which is also concurrently cost-effective. Therefore, the expression of miRNAs plays a significant role in differentiating cancer from normal, suggesting its role as a future diagnostic, prognostic, and predictive biomarker for cancer therapy. Since a single miRNA can regulate the expression of multiple target genes, it has a substantial potential for therapeutic use. Forced overexpression of pre-miRNA or inhibition of miRNA expression by antisense miRNA as demonstrated in mouse models could reduce tumor growth [80, 81]. These antisense miRNAs are now being tested in the clinical setting [82]. However, the use of natural agents or their synthetic analogs appears to have a great promise toward cancer prevention and therapy which indeed could be attributed to its function as the deregulators of miRNAs. Thus the use of natural agents or their derivatives as a single agent or in combination with conventional therapeutics will become a better strategy for prevention and/or treatment of human malignancies including BC. Although the understanding of miRNA function and regulation has increased in recent years, we are still in need of new ideas and techniques involving miRNA-based research especially for miRNA-targeted cancer therapy. Nevertheless, the future looks brighter for developing miRNA-targeted therapy for all human cancers including BC (Table 2.1).

References

1. Desantis C, Ma J, Bryan L, Jemal A (2013) Breast cancer statistics, 2013. CA Cancer J Clin 64 (1):52–62, PM:24114568
2. Ali AS, Ahmad A, Ali S, Bao B, Philip PA, Sarkar FH (2013) The role of cancer stem cells and miRNAs in defining the complexities of brain metastasis. J Cell Physiol 228:36–42, PM:22689345 PMC3443527
3. Sethi S, Ali S, Kong D, Philip PA, Sarkar FH (2013) Clinical implication of microRNAs in molecular pathology. Clin Lab Med 33:773–86, PM:24267185
4. Melo SA, Esteller M (2011) Dysregulation of microRNAs in cancer: playing with fire. FEBS Lett 585:2087–99, PM:20708002
5. Ali S, Almhanna K, Chen W, Philip PA, Sarkar FH (2010) Differentially expressed miRNAs in the plasma may provide a molecular signature for aggressive pancreatic cancer. Am J Transl Res 3:28–47, PM:21139804 PMC2981424
6. Ali S, Saleh H, Sethi S, Sarkar FH, Philip PA (2012) MicroRNA profiling of diagnostic needle aspirates from patients with pancreatic cancer. Br J Cancer 107:1354–1360, PM:22929886 PMC3494446
7. Iorio MV, Ferracin M, Liu CG, Veronese A, Spizzo R, Sabbioni S et al (2005) MicroRNA gene expression deregulation in human breast cancer. Cancer Res 65:7065–7070, PM:16103053
8. Liu J, Mao Q, Liu Y, Hao X, Zhang S, Zhang J (2013) Analysis of miR-205 and miR-155 expression in the blood of breast cancer patients. Chin J Cancer Res 25:46–54, PM:23372341 PMC3555294
9. Peng X, Guo W, Liu T, Wang X, Tu X, Xiong D et al (2011) Identification of miRs-143 and -145 that is associated with bone metastasis of prostate cancer and involved in the regulation of EMT. PLoS One 6:e20341, PM:21647377 PMC3103579
10. Wang Y, Gu J, Roth JA, Hildebrandt MA, Lippman SM, Ye Y et al (2013) Pathway-based serum microRNA profiling and survival in patients with advanced stage non-small cell lung cancer. Cancer Res 73:4801–4809, PM:23774211 PMC3760306
11. Ferracin M, Querzoli P, Calin GA, Negrini M (2011) MicroRNAs: toward the clinic for breast cancer patients. Semin Oncol 38:764–775, PM:22082762
12. Biragyn A, Bodogai M, Olkhanud PB, Denny-Brown SR, Puri N, Ayukawa K et al (2013) Inhibition of lung metastasis by chemokine CCL17-mediated in vivo silencing of genes in CCR4+ Tregs. J Immunother 36:258–267, PM:23603860 PMC3707614
13. Zhu W, Qin W, Atasoy U, Sauter ER (2009) Circulating microRNAs in breast cancer and healthy subjects. BMC Res Notes 2:89, PM:19454029 PMC2694820
14. Zhao H, Shen J, Medico L, Wang D, Ambrosone CB, Liu S (2010) A pilot study of circulating miRNAs as potential biomarkers of early stage breast cancer. PLoS One 5:e13735, PM:21060830 PMC2966402
15. Sethi S, Ali S, Kong D, Philip PA, Sarkar FH (2013) Clinical implication of microRNAs in molecular pathology. Clin Lab Med 33:773–786, PM:24267185
16. Czyzyk-Krzeska MF, Zhang X (2014) MiR-155 at the heart of oncogenic pathways. Oncogene 33(6):677–678, PM:23416982
17. Orso F, Balzac F, Marino M, Lembo A, Retta SF, Taverna D (2013) miR-21 coordinates tumor growth and modulates KRIT1 levels. Biochem Biophys Res Commun 438:90–96, PM:23872064 PMC3750217
18. Piva R, Spandidos DA, Gambari R (2013) From microRNA functions to microRNA therapeutics: novel targets and novel drugs in breast cancer research and treatment (Review). Int J Oncol 43:985–994, PM:23939688 PMC3829774
19. Sun X, Qin S, Fan C, Xu C, Du N, Ren H (2013) Let-7: a regulator of the ERalpha signaling pathway in human breast tumors and breast cancer stem cells. Oncol Rep 29:2079–2087, PM:23467929
20. Liang Y, Ridzon D, Wong L, Chen C (2007) Characterization of microRNA expression profiles in normal human tissues. BMC Genomics 8:166, PM:17565689 PMC1904203

21. Gall TM, Frampton AE, Krell J, Castellano L, Stebbing J, Jiao LR (2013) Blood-based miRNAs as noninvasive diagnostic and surrogative biomarkers in colorectal cancer. Expert Rev Mol Diagn 13:141–145, PM:23477554

22. Si H, Sun X, Chen Y, Cao Y, Chen S, Wang H et al (2013) Circulating microRNA-92a and microRNA-21 as novel minimally invasive biomarkers for primary breast cancer. J Cancer Res Clin Oncol 139:223–229, PM:23052693 PMC3549412

23. Chen Y, Gelfond JA, McManus LM, Shireman PK (2009) Reproducibility of quantitative RT-PCR array in miRNA expression profiling and comparison with microarray analysis. BMC Genomics 10:407, PM:19715577 PMC2753550

24. Perou CM, Sorlie T, Eisen MB, van de Rijn M, Jeffrey SS, Rees CA et al (2000) Molecular portraits of human breast tumours. Nature 406:747–752, PM:10963602

25. Hedenfalk I, Duggan D, Chen Y, Radmacher M, Bittner M, Simon R et al (2001) Gene-expression profiles in hereditary breast cancer. N Engl J Med 344:539–548, PM:11207349

26. Tang P, Skinner KA, Hicks DG (2009) Molecular classification of breast carcinomas by immunohistochemical analysis: are we ready? Diagn Mol Pathol 18:125–132, PM:19704256

27. Gruver AM, Portier BP, Tubbs RR (2011) Molecular pathology of breast cancer: the journey from traditional practice toward embracing the complexity of a molecular classification. Arch Pathol Lab Med 135:544–557, PM:21526953

28. Kulshreshtha R, Ferracin M, Wojcik SE, Garzon R, Alder H, Agosto-Perez FJ et al (2007) A microRNA signature of hypoxia. Mol Cell Biol 27:1859–1867, PM:17194750 PMC1820461

29. Bloomston M, Frankel WL, Petrocca F, Volinia S, Alder H, Hagan JP et al (2007) MicroRNA expression patterns to differentiate pancreatic adenocarcinoma from normal pancreas and chronic pancreatitis. JAMA 297:1901–1908, PM:17473300

30. Gottardo F, Liu CG, Ferracin M, Calin GA, Fassan M, Bassi P et al (2007) Micro-RNA profiling in kidney and bladder cancers. Urol Oncol 25:387–392, PM:17826655

31. Yanaihara N, Caplen N, Bowman E, Seike M, Kumamoto K, Yi M et al (2006) Unique microRNA molecular profiles in lung cancer diagnosis and prognosis. Cancer Cell 9:189–198, PM:16530703

32. Feber A, Xi L, Luketich JD, Pennathur A, Landreneau RJ, Wu M et al (2008) MicroRNA expression profiles of esophageal cancer. J Thorac Cardiovasc Surg 135:255–260, PM:18242245 PMC2265073

33. Ichimi T, Enokida H, Okuno Y, Kunimoto R, Chiyomaru T, Kawamoto K et al (2009) Identification of novel microRNA targets based on microRNA signatures in bladder cancer. Int J Cancer 125:345–352, PM:19378336

34. Sempere LF, Christensen M, Silahtaroglu A, Bak M, Heath CV, Schwartz G et al (2007) Altered MicroRNA expression confined to specific epithelial cell subpopulations in breast cancer. Cancer Res 67:11612–11620, PM:18089790

35. Nassirpour R, Mehta PP, Baxi SM, Yin MJ (2013) miR-221 promotes tumorigenesis in human triple negative breast cancer cells. PLoS One 8:e62170, PM:23637992 PMC3634767

36. Qi L, Bart J, Tan LP, Platteel I, Sluis T, Huitema S et al (2009) Expression of miR-21 and its targets (PTEN, PDCD4, TM1) in flat epithelial atypia of the breast in relation to ductal carcinoma in situ and invasive carcinoma. BMC Cancer 9:163, PM:19473551 PMC2695476

37. Ali S, Ahmad A, Banerjee S, Padhye S, Dominiak K, Schaffert JM et al (2010) Gemcitabine sensitivity can be induced in pancreatic cancer cells through modulation of miR-200 and miR-21 expression by curcumin or its analogue CDF. Cancer Res 70:3606–3617, PM:20388782 PMC2978024

38. Bao B, Ali S, Kong D, Sarkar SH, Wang Z, Banerjee S et al (2011) Anti-tumor activity of a novel compound-CDF is mediated by regulating miR-21, miR-200, and PTEN in pancreatic cancer. PLoS One 6:e17850, PM:21408027 PMC3052388

39. Roy S, Yu Y, Padhye SB, Sarkar FH, Majumdar AP (2013) Difluorinated-curcumin (CDF) restores PTEN expression in colon cancer cells by down-regulating miR-21. PLoS One 8: e68543, PM:23894315 PMC3722247

40. Chen J, Wang X (2014) MicroRNA-21 in breast cancer: diagnostic and prognostic potential. Clin Transl Oncol 16(3):225–233, PM:24248894

41. Si ML, Zhu S, Wu H, Lu Z, Wu F, Mo YY (2007) miR-21-mediated tumor growth. Oncogene 26:2799–2803, PM:17072344

42. Chen L, Li Y, Fu Y, Peng J, Mo MH, Stamatakos M et al (2013) Role of deregulated microRNAs in breast cancer progression using FFPE tissue. PLoS One 8:e54213, PM:23372687 PMC3553092

43. Kong W, He L, Coppola M, Guo J, Esposito NN, Coppola D et al (2010) MicroRNA-155 regulates cell survival, growth, and chemosensitivity by targeting FOXO3a in breast cancer. J Biol Chem 285:17869–17879, PM:20371610 PMC2878550

44. O'Day E, Lal A (2010) MicroRNAs and their target gene networks in breast cancer. Breast Cancer Res 12:201, PM:20346098 PMC2879559

45. Zheng SR, Guo GL, Zhai Q, Zou ZY, Zhang W (2013) Effects of miR-155 antisense oligonucleotide on breast carcinoma cell line MDA-MB-157 and implanted tumors. Asian Pac J Cancer Prev 14:2361–2366, PM:23725141

46. Ciafre SA, Galardi S, Mangiola A, Ferracin M, Liu CG, Sabatino G et al (2005) Extensive modulation of a set of microRNAs in primary glioblastoma. Biochem Biophys Res Commun 334:1351–1358, PM:16039986

47. Ma L, Teruya-Feldstein J, Weinberg RA (2007) Tumour invasion and metastasis initiated by microRNA-10b in breast cancer. Nature 449:682–688, PM:17898713

48. Tan HX, Wang Q, Chen LZ, Huang XH, Chen JS, Fu XH et al (2010) MicroRNA-9 reduces cell invasion and E-cadherin secretion in SK-Hep-1 cell. Med Oncol 27:654–660, PM:19572217

49. Chan M, Liaw CS, Ji SM, Tan HH, Wong CY, Thike AA et al (2013) Identification of circulating microRNA signatures for breast cancer detection. Clin Cancer Res 19:4477–4487, PM:23797906

50. Zhao G, Guo J, Li D, Jia C, Yin W, Sun R et al (2013) MicroRNA-34a suppresses cell proliferation by targeting LMTK3 in human breast cancer MCF-7 cell line. DNA Cell Biol 32:699–707, PM:24050776 PMC3864372

51. Guo J, Li W, Shi H, Xie X, Li L, Tang H et al (2013) Synergistic effects of curcumin with emodin against the proliferation and invasion of breast cancer cells through upregulation of miR-34a. Mol Cell Biochem 382:103–111, PM:23771315

52. Li XJ, Ji MH, Zhong SL, Zha QB, Xu JJ, Zhao JH et al (2012) MicroRNA-34a modulates chemosensitivity of breast cancer cells to adriamycin by targeting Notch1. Arch Med Res 43:514–521, PM:23085450

53. Nugent M, Miller N, Kerin MJ (2012) Circulating miR-34a levels are reduced in colorectal cancer. J Surg Oncol 106:947–952, PM:22648208

54. Wang S, Huang J, Lyu H, Lee CK, Tan J, Wang J et al (2013) Functional cooperation of miR-125a, miR-125b, and miR-205 in entinostat-induced downregulation of erbB2/erbB3 and apoptosis in breast cancer cells. Cell Death Dis 4:e556, PM:23519125 PMC3615747

55. Scott GK, Goga A, Bhaumik D, Berger CE, Sullivan CS, Benz CC (2007) Coordinate suppression of ERBB2 and ERBB3 by enforced expression of micro-RNA miR-125a or miR-125b. J Biol Chem 282:1479–1486, PM:17110380

56. Guo X, Wu Y, Hartley RS (2009) MicroRNA-125a represses cell growth by targeting HuR in breast cancer. RNA Biol 6:575–583, PM:19875930 PMC3645467

57. Ahmad A, Aboukameel A, Kong D, Wang Z, Sethi S, Chen W et al (2011) Phosphoglucose isomerase/autocrine motility factor mediates epithelial-mesenchymal transition regulated by miR-200 in breast cancer cells. Cancer Res 71:3400–3409, PM:21389093 PMC3085607

58. Creighton CJ, Gibbons DL, Kurie JM (2013) The role of epithelial-mesenchymal transition programming in invasion and metastasis: a clinical perspective. Cancer Manag Res 5:187–195, PM:23986650 PMC3754282

59. Kong D, Banerjee S, Ahmad A, Li Y, Wang Z, Sethi S et al (2010) Epithelial to mesenchymal transition is mechanistically linked with stem cell signatures in prostate cancer cells. PLoS One 5:e12445, PM:20805998 PMC2929211

60. Li Y, VandenBoom TG, Kong D, Wang Z, Ali S, Philip PA et al (2009) Up-regulation of miR-200 and let-7 by natural agents leads to the reversal of epithelial-to-mesenchymal transition in gemcitabine-resistant pancreatic cancer cells. Cancer Res 69:6704–6712, PM:19654291 PMC2727571

61. Park SM, Gaur AB, Lengyel E, Peter ME (2008) The miR-200 family determines the epithelial phenotype of cancer cells by targeting the E-cadherin repressors ZEB1 and ZEB2. Genes Dev 22:894–907, PM:18381893 PMC2279201

62. Li X, Roslan S, Johnstone CN, Wright JA, Bracken CP, Anderson M et al (2013) MiR-200 can repress breast cancer metastasis through ZEB1-independent but moesin-dependent pathways. Oncogene. doi:10.1038/onc.2013.370, PM:24037528

63. Manavalan TT, Teng Y, Litchfield LM, Muluhngwi P, Al-Rayyan N, Klinge CM (2013) Reduced expression of miR-200 family members contributes to antiestrogen resistance in LY2 human breast cancer cells. PLoS One 8:e62334, PM:23626803 PMC3633860

64. Bai JX, Yan B, Zhao ZN, Xiao X, Qin WW, Zhang R et al (2013) Tamoxifen represses miR-200 microRNAs and promotes epithelial-to-mesenchymal transition by up-regulating c-Myc in endometrial carcinoma cell lines. Endocrinology 154:635–645, PM:23295740

65. Lim YY, Wright JA, Attema JL, Gregory PA, Bert AG, Smith E et al (2013) Epigenetic modulation of the miR-200 family is associated with transition to a breast cancer stem-cell-like state. J Cell Sci 126:2256–2266, PM:23525011

66. Li Z, Xiao J, Wu X, Li W, Yang Z, Xie J et al (2012) Plumbagin inhibits breast tumor bone metastasis and osteolysis by modulating the tumor-bone microenvironment. Curr Mol Med 12:967–981, PM:22574935

67. Ahmad A, Wang Z, Ali R, Maitah MY, Kong D, Banerjee S et al (2010) Apoptosis-inducing effect of garcinol is mediated by NF-kappaB signaling in breast cancer cells. J Cell Biochem 109:1134–1141, PM:20108249

68. Ahmad A, Wang Z, Wojewoda C, Ali R, Kong D, Maitah MY et al (2011) Garcinol-induced apoptosis in prostate and pancreatic cancer cells is mediated by NF- kappaB signaling. Front Biosci (Elite Ed) 3:1483–1492, PM:21622152

69. Ahmad A, Sarkar SH, Bitar B, Ali S, Aboukameel A, Sethi S et al (2012) Garcinol regulates EMT and Wnt signaling pathways in vitro and in vivo, leading to anticancer activity against breast cancer cells. Mol Cancer Ther 11:2193–2201, PM:22821148 PMC3836047

70. Chen CS, Lee CH, Hsieh CD, Ho CT, Pan MH, Huang CS et al (2011) Nicotine-induced human breast cancer cell proliferation attenuated by garcinol through down-regulation of the nicotinic receptor and cyclin D3 proteins. Breast Cancer Res Treat 125:73–87, PM:20229177

71. Cheng AC, Tsai ML, Liu CM, Lee MF, Nagabhushanam K, Ho CT et al (2010) Garcinol inhibits cell growth in hepatocellular carcinoma Hep3B cells through induction of ROS-dependent apoptosis. Food Funct 1:301–307, PM:21776480

72. K A T, T R, G R, K C S, Nair RS, G S et al (2013) Structure activity relationship of plumbagin in BRCA1 related cancer cells. Mol Carcinog 52:392–403, PM:22290577

73. Lee JH, Yeon JH, Kim H, Roh W, Chae J, Park HO et al (2012) The natural anticancer agent plumbagin induces potent cytotoxicity in MCF-7 human breast cancer cells by inhibiting a PI-5 kinase for ROS generation. PLoS One 7:e45023, PM:23028742 PMC3441601

74. Manu KA, Shanmugam MK, Rajendran P, Li F, Ramachandran L, Hay HS et al (2011) Plumbagin inhibits invasion and migration of breast and gastric cancer cells by downregulating the expression of chemokine receptor CXCR4. Mol Cancer 10:107, PM:21880153 PMC3175200

75. Masuelli L, Benvenuto M, Fantini M, Marzocchella L, Sacchetti P, Di SE et al (2013) Curcumin induces apoptosis in breast cancer cell lines and delays the growth of mammary tumors in neu transgenic mice. J Biol Regul Homeost Agents 27:105–119, PM:23489691

76. Sun SH, Huang HC, Huang C, Lin JK (2012) Cycle arrest and apoptosis in MDA-MB-231/ Her2 cells induced by curcumin. Eur J Pharmacol 690:22–30, PM:22705896
77. Shehzad A, Khan S, Sup LY (2012) Curcumin molecular targets in obesity and obesity-related cancers. Future Oncol 8:179–190, PM:22335582
78. Yodkeeree S, Ampasavate C, Sung B, Aggarwal BB, Limtrakul P (2010) Demethoxycurcumin suppresses migration and invasion of MDA-MB-231 human breast cancer cell line. Eur J Pharmacol 627:8–15, PM:19818349
79. Kanwar SS, Yu Y, Nautiyal J, Patel BB, Padhye S, Sarkar FH et al (2011) Difluorinated-curcumin (CDF): a novel curcumin analog is a potent inhibitor of colon cancer stem-like cells. Pharm Res 28:827–838, PM:21161336 PMC3792588
80. Esau C, Davis S, Murray SF, Yu XX, Pandey SK, Pear M et al (2006) miR-122 regulation of lipid metabolism revealed by in vivo antisense targeting. Cell Metab 3:87–98, PM:16459310
81. Kota J, Chivukula RR, O'Donnell KA, Wentzel EA, Montgomery CL, Hwang HW et al (2009) Therapeutic microRNA delivery suppresses tumorigenesis in a murine liver cancer model. Cell 137:1005–1017, PM:19524505 PMC2722880
82. Nana-Sinkam SP, Croce CM (2013) Clinical applications for microRNAs in cancer. Clin Pharmacol Ther 93:98–104, PM:23212103

Chapter 3
Epidemiology, Risk Factors, Treatment, and Prevention of Breast Cancer Metastases

Manal Nizam, Saba Haq, Shadan Ali, Raagini Suresh,
Ramzi M. Mohammad, and Fazlul H. Sarkar

Abstract A large percentage of Americans are plagued by cancer. In the United States, one third of females and one half of males face the condition sometime in their lifetime. One type of cancer that has adversely impacted the lives of many on a multinational scale is breast cancer (BC). In order to effectively battle and prevent the condition of BC, it is critical to obtain a better understanding of the biology of BC and its metastases. An important step to gain better understanding is by studying risk factors. Under this category, it is important to consider both occupational hazards and genetic predispositions. Additionally, drugs are ever-important in the fight against any disease. In the fight against BC, it is important to be aware of new drugs for BC prevention and cure. However, older drugs should not be disregarded, and thus, studies should continue to search for new information concerning their side effects and uses. Lastly, since 90 % of cancer deaths occur due to metastases, this aspect of cancer cannot be disregarded. Prevention and treatment methods for BC metastases must also be considered. Thus, this article will represent discussion on risk factors, information about new and existent drugs, and treatment and prevention for BC metastases, especially to bones and the brain. This article will also discuss new ways to synergize existing conventional drugs and preventive technologies in order to achieve optimal management strategies for eradicating BC.

Keywords Breast cancer • Cancer epidemiology • Prevention • Treatment • Metastases

M. Nizam • S. Haq • R. Suresh
Department of Pathology, Karmanos Cancer Institute, Wayne State University School of Medicine, 740 Hudson Webber Cancer Research Center, 4100 John R Street, Detroit, MI, USA

S. Ali • R.M. Mohammad
Department of Oncology, Karmanos Cancer Institute, Wayne State University School of Medicine, Detroit, MI, USA

F.H. Sarkar (✉)
Department of Pathology, Karmanos Cancer Institute, Wayne State University School of Medicine, 740 Hudson Webber Cancer Research Center, 4100 John R Street, Detroit, MI, USA

Department of Oncology, Karmanos Cancer Institute, Wayne State University School of Medicine, Detroit, MI, USA
e-mail: fsarkar@med.wayne.edu

© Springer International Publishing Switzerland 2014   23
S. Sethi, *miRNAs and Target Genes in Breast Cancer Metastasis*,
SpringerBriefs in Cancer Research, DOI 10.1007/978-3-319-08162-5_3

Introduction

In attempts to thwart disease, it is often effective to employ prevention measures, including the implementation of frequent screening strategies. Before such methods are possible for any disease or condition, that condition's underlying epidemiology must be understood. In the fight against breast cancer (BC), steps toward understanding its epidemiology include identifying causes and risk factors for this condition. With a heightened understanding of these aspects, the most useful screening methods can be determined.

Genetic factors certainly play a role in causing these types of cancer. In combination with environmental risk factors, genetic mutations and hereditary genes can have significantly adverse impacts. The interaction between genetic factors and environmental risk factors can be seen when looking at the methylenetetrahydrofolate reductase (MTHFR) genotype. Factors such as cigarette and alcohol usage synergistically act with the MTHFR genotype to break down DNA strands [1]. In addition to genetics and substance abuse, high body mass indexes (BMIs) and weight gain increase patients' susceptibilities to cancer [2]. Identifying risk factors and linkages between them is critical because they inform patients how to change their lifestyles to lower their chances of developing cancer.

Thankfully, previously conducted research has allowed for the development of cancer prevention techniques. One such example of a preventive technique is vitamin intake. Vitamins that are successful at preventing cancer include vitamin D, vitamin C, calcium, and vitamin B_c [3–6]. These vitamins help suppress the growth of tumors and can help eradicate potentially malignant abnormal cells. Furthermore, effective screening practices have been shown to lower cancer risk. Screening techniques for BC include mammography and MRI [7, 8]. Prophylactic surgery, chemoprophylaxis, and aromatase inhibitors can also be used to reduce mortality of high-risk patients [7, 9–11]. As always, these prevention and screening techniques have side effects that must be taken into account and carefully considered before their implementation.

This article will discuss BC epidemiology in depth, delving into the causes of this cancer, as well as relevant prevention and screening techniques. Some of the discussed strategies are still being researched and developed. The continuous development of such techniques leads to new discoveries, contributing to a hopeful future of even more effective cancer prevention.

Breast Cancer

Breast cancer (BC) is the cancer that most frequently results in mortality for women [12]. More than three million women across the United States have been diagnosed with it at one point in time, usually around the age of 61 [13]. The conception of BC can be caused by genetic factors or by environmental influences. BC, hereditary or

Table 3.1 Differences between BRCA 1 and BRCA 2 mutations

BRCA 1	BRCA 2	References
• 80 % of BC have been in premenopausal stage women with the BRCA 1 mutation • BRCA 1 carriers have an earlier onset of BC than those who carry a BRCA 2 mutation • BRCA 1 mutation gives a high risk of ovarian cancer	• BRCA 2 mutation carriers usually develop postmenopausal BC, though they also have a significant risk of developing BC before menopause	[7, 9]

sporadic, can be effectively prevented by understanding the underlying epidemiology, assessing risk factors, and utilizing aggressive prevention and/or even treatment strategies.

Hereditary Breast Cancer

Hereditary BC is usually the result of mutations in the BRCA 1 and BRCA 2 genes, which are present in 2–6 % of all those who develop BC [7] in their lifetime. The carriers of BRCA 1/2 mutations are more likely to lead to the development of BC than are noncarriers, with 40–80 % of the carriers acquiring BC in their lifetimes [7]. The differences between BRCA 1 and BRCA 2 mutations are depicted in Table 3.1. BRCA 1 mutation carriers typically have an earlier onset of BC than BRCA 2 mutation carriers. In fact, approximately 80 % of BC occurrences in BRCA 1 mutation carriers are in the premenopausal stage [7]. On the other hand, BRCA 2 mutation carriers usually develop postmenopausal BC, though they also have a significant risk of developing BC before menopause [9].

BRCA 1/2 mutations are extremely detrimental, which can be attributed to the function of the BRCA 1/2 genes. The BRCA 1/2 genes participate in active DNA repair of double-strand break. Furthermore, the repair processes of these genes are predisposed to being erroneous [7]. If the BRCA 1/2 genes are mutated, DNA repair is carried out incorrectly, leading to genetic miscoding. This causes the cells to start the process of carcinogenesis [7].

These mutations also have a pleiotropic effect, as carriers of BRCA 1/2 are more susceptible to other cancers, especially ovarian cancer [9]. One study states that BRCA 1 mutation carriers have equally high risks of acquiring ovarian cancer and BC [9]. The Manchester scoring model can be used to assess the risk of possessing these harmful mutations [7]. This model is extremely sensitive, and it takes family medical history of cancer into account as well as the age for BC initiation and development. A drawback to this model is that some believe it over-refers patients for genetic tests [7]. However, the US Preventive Services Task Force (USPSTF) recommends that anyone with a family history of BC should get screened for BRCA 1/2 mutations [14].

Smoking and Breast Cancer

Cigarettes give off tobacco smoke that contains carcinogens, such as polycyclic aromatic hydrocarbons, aromatic amines, and N-nitrosamines, which are transmitted through the alveolar membrane to the blood stream and then straight to mammary tissue which can pose detrimental effects on health [15]. Since ever-smokers are more exposed to these carcinogens than past, current, or never-smokers, they are at a higher risk for developing BC [15]. Furthermore, there is a higher associated risk for women who have smoked before first birth and before menarche [15, 16]. Additionally, the quantity of cigarettes smoked has been positively correlated with BC risk [15]. People who started smoking before they were 18 years of age, had been smoking since then for more than 35 years, and had an intake of more than 25 cigarettes a day had a 125 % increased risk of BC than those who never smoked [15]. Thus, cigarette smoking should be avoided to prevent the risk of developing BC, in addition to many other human malignancies.

BMI, Obesity, and Breast Cancer

More than 20 % of breast tumors in the United States can be, at least, partially explained by weight gain [2]. Body mass index (BMI) has been positively correlated with obesity; thus, as BMI and obesity increase, BC risk in premenopausal women goes up. BMI has also been correlated with mammographic density (MD), which is a measure of the density of breast tissue and is a marker used to assess BC risks. The relationship between MD and BC is inversely proportional because the breast is primarily made up of adipose tissue, which is mostly composed of fat storage sites [2]. MD alters with age, BMI, and with time due to menopause [2]. Hence, it is important for women to maintain a healthy BMI, MD, and weight, as this may help to lower the risk for BC.

Oral Contraceptives

Long-term usage of oral contraceptives has been known to greatly increase the risk of developing BC [9, 10, 17]. Women who started using oral contraceptives at an early stage of life, around 23–27 years of age, are more likely to have early onset of BC, especially if they are carriers of the BRCA 1 mutation. However, oral contraceptives have been shown to reduce ovarian cancer occurrences [9, 10]. To be faced with an increased risk of BC from the usage of oral contraceptives, the user has to have either used them in the past 10 years and/or is currently using them for increased risk of BC [17].

Oral contraceptives contain progestin and estrogen, a combination that causes an increased risk of BC conception in postmenopausal women. This increased risk is due to the increased breast cell proliferation associated with higher levels of these hormones [17]. Oral contraceptives also indirectly lead users to be associated with other risk factors, including fewer births and infrequent breast-feeding [17]. Hence, it is essential that users of oral contraceptives undergo frequent screening for early detection and diagnosis of BC.

Vitamin D and Its Correlation with Breast Cancer

Vitamin D exhibits many properties that inhibit the development of carcinogenesis, which include decrease in angiogenesis, increase in apoptosis, and maintenance of breast cells. Vitamin D metabolites such as serum 25-hydroxyvitamin D [1,25 $(OH)_2D$] also help maintain breast cells by inhibiting the production of COX-2 enzymes, which induce angiogenesis [5]. Sources of vitamin D include ultraviolet B rays, oral vitamin D supplements, and the presence of 25-hydroxyvitamin D in the serum [25(OH)D]. Vitamin D helps regulate the adhesion between cells in breast epithelial tissue, thereby helping to prevent the overgrowth of cells. Thus, it is important to increase vitamin D intake to protect women from the development of malignancy. Research shows that there is a reduction of BC risk of up to 80 % if serum 25(OH)D levels are increased [5]. Studies have also demonstrated that BC diagnosis increases during the winter season, due to lower serum levels which are at its lowest, indicating a positive role of sun-induced vitamin D in the prevention of BC [5]. Such research brings to light an important preventive technique; individuals should ensure their vitamin D intakes are at suitable levels to help protect them from BC.

Screening

There is a correlation between better BC detection and prevention. Individuals with more frequent primary care doctor visits are more likely to get mammographies done, which help to screen for BC [8]. There do exist, however, certain limitations to mammography. For example, it is difficult to detect smaller lumps in breast tissue that has become dense due to the dense parenchyma acquired prior to menopause [18].

Mammography is a preventive screening method that many believe should be used annually, though there do exist some minor disagreements [7]. The USPSTF recommends that women between the ages of 50 and 74 get a mammogram, at least, once every 2 years as presented in Table 3.2. In addition, it is recommended that women have a physical breast examination every 6–12 months or routine self-examination on a daily basis to help identify tumors that are smaller than 1 cm in

Table 3.2 Breast cancer screening methods

Screening methods	Recommendation/usage	References
Mammography	Should be used annually to screen for women between 50 and 74	[7]
Physical breast examination	Should be done every 6–12 months to help detect large tumors	[7, 18]
MRI	Helps detect later stages of breast cancer (BC) and is good for detecting BC in BRCA 1 carriers	[9]
Ductal lavage and periareolar fine needle aspirate	Uses cytological atypia in the breast fluid or the increase of breast aspirate to assess the risk of developing BC	[19]

diameter [7, 18]. For women with a high risk for BC, MRIs can also be performed for more thorough screening. MRIs can be performed as early as the age of 25, depending on the patient's family history [7].

MRI can show tumors that go undetected by mammography [9]. However, MRI is only sensitive enough to detect later stages of BC [9]. Its advantages lie in the fact that, unlike mammography, it is good at detecting tumors in BRCA 1 carriers. Approximately half of all BRCA mutations found in patients are detected solely through the use of MRI screening [9]. Unfortunately, this intensified sensitivity leads to many unnecessary biopsies [7]. Research has shown that combining findings of mammography and MRI helps to reduce these unneeded biopsies [7]. The sensitivity range for the two procedures together is 93–100 %, while mammography alone gives only a 47–67 % sensitivity range [7]. Such screening methods are invaluable in helping healthcare professionals strive to detect BC in its earlier stages where management will lead to cure.

In addition to the aforementioned screening methods, there are alternative and newer methods that can also be used to assess risk. These include random periareolar fine needle aspirate (RPFNA) and ductal lavage (DL) that uses cytological atypia in the breast fluid or the increase of breast aspirate to assess the risk of developing BC [19]. RPFNA is a test that is inexpensive, repeatable and can assess the short-term risk of developing BC. If atypical hyperplasia is detected using RPFNA, tamoxifen could be given to reduce the risk of BC, because atypical hyperplasia and BC are positively correlated [19].

Ductal lavage is an experimental technology that assesses nipple fluid to find malignant cells from the lesions. This process helps evaluate the risk of BC and aids in identifying the location of any abnormal cells [19]. Its primary disadvantage lies in the fact that it has been reported 51 % less comfortable than mammography. In fact, 44 % of patients who had undergone DL reported breast pain. The specificity of this technology is 79 %, and its sensitivity is 47 % [19]. Ductal lavage is used infrequently since its value is still under debate.

Prophylactic Surgery

The use of prophylactic surgery as a preventive measure is advised to those who are at the higher-risk category for the development of BC, especially for carriers of the BRCA 1/2 mutations [10]. The prophylactic mastectomy procedure is known to reduce the occurrence of BC. It can add as many as 11.7 years to a patient's life [11]. Another procedure is the bilateral salpingo-oophorectomy which can reduce the chances of developing BC by 80 % [7]. This procedure, like prophylactic mastectomy, is recommended for subjects at the age of 40, or around the end of a woman's childbearing age, to improve survival in those with BRCA mutations. Statistical data show a 24 % increase in the survival rate for women with BRCA 1 mutations who have undergone these procedures and an 11 % increase in the survival rate for women with BRCA 2 mutations [7]. Thus, prophylactic surgery has been proven to be an effective preventive measure.

Chemoprophylaxis in Breast Cancer

Selective Estrogen Receptor Modulators

Selective estrogen receptor modulators (SERMs) are typically used to lower the risk of BC in high-risk patients. Examples of SERMs include tamoxifen, raloxifene, and exemestane [11]. Tamoxifen is the SERM most commonly used by patients and is beneficial for BRCA 1 mutation carriers also, because it reduces their risk of contralateral BC [9]. Chemoprophylaxis is also useful in treating patients with BRCA 2 mutations, as it decreases their BC risk by 62 %, according to the National Surgical Adjuvant Breast and Bowel Project [7, 11, 20]. In addition, chemoprophylaxis has been shown to increase the life span of an individual at high risk for BC by 1.6 years [11]. One caveat with tamoxifen use is that it has been known to have an increased risk of endometrial cancer, and although the risk is very low, the drug's usage should be limited to those at highest risk for BC [7].

The drug raloxifene is primarily used to prevent osteoporosis, but it is also a preventive method for high-risk BC patients [7]. In fact, it has been known to increase the average life span by 2.2 years [11]. This drug has the same efficacy as tamoxifen in terms of preventing invasive BC. Though it does not increase the risk of developing uterine cancer as much as tamoxifen does, raloxifene users acquire an increased risk of developing noninvasive BC [7].

Exemestane is another SERM that is used to prevent BC in women who are postmenopausal. The drug increases estrogen by excessively stimulating the ovaries. Exemestane is not as beneficial of a preventive treatment as tamoxifen and is only used in women who are not at a risk for uterine cancer [20].

Aromatase Inhibitors

Aromatase inhibitors are estrogen production inhibiting agents and are used as a preventive method against BC. They are known to be more active at preventing ER-positive BC [9, 11] and also more effective than tamoxifen in preventing contralateral BC in women who are postmenopausal [11].

Prevention of Breast Cancer with Retinoids

Vitamin A

Retinoids are derivatives of Vitamin A and are an effective method of BC prevention. The most popularly used retinoid is fenretinide, which suppresses tumor growth. Fenretinide can also change the genes that maintain the tumor and cause apoptosis of cells in the tumor by inhibiting growth signaling [3, 4]. It is recommended that fenretinide be used by women who are younger than 40 years of age because its usage can decrease BC risk by 50 % [3]. Vitamin A is especially beneficial for women who fall in the high risk of BC category, including those who possess BRCA 1 mutation [3]. Due to its beneficial properties, getting the recommended intake of vitamin A is critical, especially for women with high risk of developing BC.

Vitamin E

Vitamin E in the form of tocopherols is another advantageous preventive method when administered to patients with high risk of BC. This is especially true for γ- and δ-tocopherol, because they inhibit estrogen signaling and reduce cell proliferation in ER-positive BC [6]. The γ-tocopherol has also been shown to reduce mammary hyperplasia and reduce the occurrences of tumors [6].

Introduction to Breast Cancer Metastases

The major cause of death from BC includes metastases that arise in the lung, bone, liver, and brain [21, 22]. The 5-year survival rate is only 26 % for those affected with metastatic breast cancer [22]. Bone metastases can also arise 10 years after the primary tumor is eradicated, since malignant cells take time to have secondary outgrowth [23]. Alarmingly, metastases are the cause for almost 90 % of all cancer deaths [21]. Currently, the understanding of the cure and prevention of cancer metastases is extraordinarily poor, which brings it to the forefront of issues needing further research.

Metastases arise when the cancer cells leave the primary tumor, called circulating tumor cells (CTCs); enter surrounding tissues; start circulating in the vasculature; and start the process of extravasation [21]. This process of metastasis involves the CTCs establishing transient and reaching the secondary sites by crossing the endothelial and pericyte layers. Metastasis commencement is dependent upon the presence of specific proteins [21]. The most vital protein needed for metastasis to arise is the MMP-2 protein [22].

The two existing models for understanding metastasis are clonal evolution and cancer stem cell theory [24]. Clonal evolution theory states that cancer cells have multiple genetic alterations that result in different processes of metastasis and tumor growth. Cancer stem cell theory states that tumor cells are not all alike due to their hierarchal organization [24]. At the top of the hierarchy lie cancer stem cells. This theory also establishes the idea that there are certain changes in gene expression that lead to metastasis, which is what drives the cellular events that are responsible for tumor growth and metastasis. By studying these genes, patterns can be identified, which would likely lead to gain further information about metastasis, and such knowledge could be useful for improving treatment strategies for the prevention and/or treatment of BC metastasis [24].

Bone Metastasis

Since bone is the site for 30–40 % of first tumor reoccurrences, bone metastases are the most common metastases for BC patients [25]. The bone has a host of growth factors, and patients with bone metastases have better prognoses than patients with visceral metastasis to the brain. This is because estrogen and progesterone receptors are absent in visceral metastasis [25]. On the other hand, bone metastasis occurs most frequently with estrogen receptor-positive tumors and can happen any time between 10 and 15 years after curative treatments [25]. Bone metastasis occurs in 70 % of advanced BC patients. It causes the compression of the spinal cord, causing fractures to develop, which can often lead to extreme pain and death [21]. It is the most common and highest mortality causing metastatic site for BC [25]. While long bone metastasis occurs very often, jaw metastasis occurs rarely, often in conjunction with advanced stage breast cancer [26].

Both in vivo and ex vivo models have been used to study the metastases of BC into the bone through the process of extravasation [21]. As discussed earlier, the process of metastasis into the bone via extravasation involves reaching secondary sites by crossing the endothelial and pericyte layers [21]. Though there are limitations in understanding the critical events that are important in these model systems, they may be useful for replicating some of the physiologic conditions. In vitro models have also been used to imitate in vivo system, by using assays such as Boyden chamber or wound healing to study cell migration by providing controlled environments [21]. One such study has developed a tri-culture microfluidic 3D in vitro model to further study extravasation by identifying the migration of metastatic BC cells and to observe their activities within the bone-like microenvironment. They observed proliferation of cancer cells that caused micrometastases within the bone cell-conditioned microenvironment following extravasation of cancer cells. These studies certainly will help to expand our knowledge of cancer biology and thus will help for screening for newer and more effective therapeutics [21].

Betulinic acid is a triterpenoid with potential use in prevention and therapy [27]. According to a recent study, the progression and intensity of not only cancer but also obesity, diabetes, and cardiovascular disease was significantly reduced by betulinic acid [27]. The effects of betulinic acid on the mouse model of BC indicated not as much loss of bone during BC metastasis, suggesting that betulinic acid may prevent bone loss in patients with bone metastases and cancer treatment-induced estrogen deficiency [27].

Brain Metastasis

Brain metastases that arise from BC are fairly common, occurring in 10–16 % of patients with advanced BC and in 15–40 % of all BC patients [28, 29]. The age range for 60 % of all breast metastases is around 50–70 years [30]. As such, brain metastasis is a major cause of mortality in BC patients. There are few options available for treatment, even though astounding amounts of research have been performed for understanding the biology of BC brain metastasis. This is a huge problem as brain metastases occur within 2–3 years after the patient is diagnosed with the onset of metastatic disease. After brain metastasis begins, the average time frame of survival is only 13 months [28]. BRCA1 germ line mutations, epidermal growth factor receptor 2 expression, younger age, ethnicity, and hormone receptor-negative status are numerous prognostic factors that have been implicated in the development of brain metastases [28].

About 80 % of the tumors in the brain occur in the cerebrum, while 15 % are found in the cerebellum, and 5 % occur in the brainstem [30]. Diagnosis of brain metastases in these regions usually involves the use of CT scans, though MRI is more specific and sensitive in detecting metastases [31]. For proper therapy, early detection is vital. CT scans and MRI are extremely valuable, since brain metastases can be either symptomatic or asymptomatic. Thus, determining which sort of metastasis is occurring in the patient is important because if neurological symptoms develop, even successful treatment cannot eradicate the symptoms [31].

Treatment for brain metastases is difficult due to the presence of the blood–brain barrier (BBB) that prevents therapeutic drugs from reaching the central nervous system [32]. Evans blue (EB) dye indicated a reasonable method of determining the status of BBB prior to euthanizing the mouse to help manage the tumors within the mouse brain parenchyma [32]. In order for the therapeutic drugs to better access the tumor, the permeability of BBB must be increased. One of these therapeutic drugs is temozolomide, which has been shown to penetrate the BBB. Though this drug alone has never been optimal in helping patients diagnosed with brain metastases, it has been demonstrated to work moderately in combination with cisplatin or capecitabine [31]. Another drug that effectively penetrates the BBB is methotrexate, but only in high-dose administration. It should be noted that this drug is not used often because it can lead to toxic leukoencephalopathy [31].

After therapeutic drugs, radiotherapy and surgical resection are the next best treatments available [32]. Radiotherapy for brain metastases can result in a prolonged survival of 4–5 months, while surgical resection can extend survival length even further [28]. In one research study, 71 % of the patients with brain metastases who had undergone treatment with radiation therapy experienced a complete disappearance of intracranial hypertension symptoms [33]. Additionally, stereotactic radiosurgery can also help prolong survival if there are less than three metastases present and all are surgically treated [28].

Conclusion

Breast cancer affects vast numbers of people. As more and more studies are conducted, new prevention techniques will continue to be unearthed. However, in order to truly fight the battle against cancer in general population, individuals must take steps to reduce their risk for developing such conditions. Preventive techniques mentioned in this paper include vitamin and mineral intakes of substances such as vitamin C, vitamin D, calcium, and vitamin B. There are also many other drugs that can help prevent the onset of breast cancer and breast cancer metastases in high-risk patients, some of which were also discussed. Another set of useful techniques in the battle against cancer can be found in risk assessment methods. These measures include evaluating occupational hazards, gender, genetic predispositions to the cancers, age, previous contraction of diseases, weight/BMI, and lifestyle risks (including the usage of alcohol and smoking).

Prevention techniques and treatment options for metastases to bones and the brain were also discussed. Betulinic acid was indicated as being a useful drug for the prevention and treatment of bone metastases. Furthermore, temozolomide and methotrexate are therapeutic drugs that can be used for brain metastases alongside treatment with radiotherapy and surgical resection. A final pivotal method that cannot be forgotten is the use of epidemiological studies that have helped and continue to help draw attention to new areas of research for further investigations for improving the lives of women around the world.

Conflict of Interest All the authors declare no competing conflict of interest.

References

1. Kobayashi LC, Limburg H, Miao Q, Woolcott C, Bedard LL, Massey TE et al (2012) Folate intake, alcohol consumption, and the methylenetetrahydrofolate reductase (MTHFR) C677T gene polymorphism: influence on prostate cancer risk and interactions. Front Oncol 2:100
2. Pollan M, Lope V, Miranda-Garcia J, Garcia M, Casanova F, Sanchez-Contador C et al (2012) Adult weight gain, fat distribution and mammographic density in Spanish pre- and post-menopausal women (DDM-Spain). Breast Cancer Res Treat 134:823–838
3. Cazzaniga M, Varricchio C, Montefrancesco C, Feroce I, Guerrieri-Gonzaga A (2012) Fenretinide (4-HPR): a preventive chance for women at genetic and familial risk? J Biomed Biotechnol 2012:172897
4. Coviello AD, Haring R, Wellons M, Vaidya D, Lehtimaki T, Keildson S et al (2012) A genome-wide association meta-analysis of circulating sex hormone-binding globulin reveals multiple loci implicated in sex steroid hormone regulation. PLoS Genet 8:e1002805
5. Mohr SB, Gorham ED, Alcaraz JE, Kane CI, Macera CA, Parsons JK et al (2012) Does the evidence for an inverse relationship between serum vitamin D status and breast cancer risk satisfy the Hill criteria? Dermatoendocrinology 4:152–157
6. Smolarek AK, Suh N (2011) Chemopreventive activity of vitamin E in breast cancer: a focus on gamma- and delta-tocopherol. Nutrients 3:962–986
7. Bougie O, Weberpals JI (2011) Clinical considerations of B. Int J Surg Oncol 2011:374012
8. Roetzheim RG, Ferrante JM, Lee JH, Chen R, Love-Jackson KM, Gonzalez EC et al (2012) Influence of primary care on breast cancer outcomes among medicare beneficiaries. Ann Fam Med 10:401–411
9. Moller P (2004) Towards evidence-based management of inherited breast and breast-ovarian cancer. Hered Cancer Clin Pract 2:11–16
10. Pasanisi P, Hedelin G, Berrino J, Chang-Claude J, Hermann S, Steel M et al (2009) Oral contraceptive use and BRCA penetrance: a case-only study. Cancer Epidemiol Biomarkers Prev 18:2107–2113
11. Salhab M, Bismohun S, Mokbel K (2010) Risk-reducing strategies for women carrying BRCA1/2 mutations with a focus on prophylactic surgery. BMC Women's Health 10:28
12. Aft RL, Naughton M, Trinkaus K, Weilbaecher K (2012) Effect of (Neo)adjuvant zoledronic acid on disease-free and overall survival in clinical stage II/III breast cancer. Br J Cancer 107:7–11
13. Siegel R, DeSantis C, Virgo K, Stein K, Mariotto A, Smith T et al (2012) Cancer treatment and survivorship statistics, 2012. CA Cancer J Clin 62:220–241
14. Nelson HD, Huffman LH, Fu R, Harris EL (2005) Genetic risk assessment and BRCA mutation testing for breast and ovarian cancer susceptibility: systematic evidence review for the U.S. Preventive Services Task Force. Ann Intern Med 143:362–379
15. Xue F, Willett WC, Rosner BA, Hankinson SE, Michels KB (2011) Cigarette smoking and the incidence of breast cancer. Arch Intern Med 171:125–133
16. Pieta B, Chmaj-Wierzchowska K, Opala T (2012) Life style and risk of development of breast and ovarian cancer. Ann Agric Environ Med 19:379–384
17. Hunter DJ, Colditz GA, Hankinson SE, Malspeis S, Spiegelman D, Chen W et al (2010) Oral contraceptive use and breast cancer: a prospective study of young women. Cancer Epidemiol Biomarkers Prev 19:2496–2502
18. Huang Y, Kang M, Li H, Li JY, Zhang JY, Liu LH et al (2012) Combined performance of physical examination, mammography, and ultrasonography for breast cancer screening among Chinese women: a follow-up study. Curr Oncol 19:eS22–eS30
19. Hoffman A, Pellenberg R, Drendall CI, Seewaldt V (2012) Comparison of random periareolar fine needle aspirate versus ductal lavage for risk assessment and prevention of breast cancer. Curr Breast Cancer Rep 4:180–187
20. Litton JK, Bevers TB, Arun BK (2012) Exemestane in the prevention setting. Ther Adv Med Oncol 4:107–112

21. Bersini S, Jeon JS, Dubini G, Arrigoni C, Chung S, Charest JL et al (2014) A microfluidic 3D in vitro model for specificity of breast cancer metastasis to bone. Biomaterials 35:2454–2461
22. Qin C, He B, Dai W, Zhang H, Wang X, Wang J et al (2014) Inhibition of metastatic tumor growth and metastasis via targeting metastatic breast cancer by chlorotoxin-modified liposomes. Mol Pharm. doi:10.1021/mp400691z
23. Bendinelli P, Maroni P, Matteucci E, Luzzati A, Perrucchini G, Desiderio MA (2014) Microenvironmental stimuli affect Endothelin-1 signaling responsible for invasiveness and osteomimicry of bone metastasis from breast cancer. Biochim Biophys Acta 1843:815–826
24. Fazilaty H, Mehdipour P (2014) Genetics of breast cancer bone metastasis: a sequential multistep pattern. Clin Exp Metastasis 31(5):595–612
25. Vona-Davis L, Rose DP, Gadiyaram V, Ducatman B, Hobbs G, Hazard H et al (2014) Breast cancer pathology, receptor status, and patterns of metastasis in a rural Appalachian population. J Cancer Epidemiol 2014:170634
26. Flores IL, Dos Santos-Silva AR, Coletta RD, Vargas PA, Lopes MA (2014) Synchronous antiresorptive osteonecrosis of the jaws and breast cancer metastasis. Oral Surg Oral Med Oral Pathol Oral Radiol 117:e264–e268
27. Park SY, Kim HJ, Kim KR, Lee SK, Lee CK, Park KK et al (2014) Betulinic acid, a bioactive pentacyclic triterpenoid, inhibits skeletal-related events induced by breast cancer bone metastases and treatment. Toxicol Appl Pharmacol 275:152–162
28. Salhia B, Kiefer J, Ross JT, Metapally R, Martinez RA, Johnson KN et al (2014) Integrated genomic and epigenomic analysis of breast cancer brain metastasis. PLoS One 9:e85448
29. Wu X, Luo B, Wei S, Luo Y, Feng Y, Xu J et al (2013) Efficiency and prognosis of whole brain irradiation combined with precise radiotherapy on triple-negative breast cancer. J Cancer Res Ther 9(Suppl):S169–S172
30. Saha A, Ghosh SK, Roy C, Choudhury KB, Chakrabarty B, Sarkar R (2013) Demographic and clinical profile of patients with brain metastases: a retrospective study. Asian J Neurosurg 8:157–161
31. Gil-Gil MJ, Martinez-Garcia M, Sierra A, Conesa G, Del BS, Gonzalez-Jimenez S et al (2014) Breast cancer brain metastases: a review of the literature and a current multidisciplinary management guideline. Clin Transl Oncol 16(5):436–446
32. Do J, Foster D, Renier C, Vogel H, Rosenblum S, Doyle TC et al (2014) Ex vivo Evans blue assessment of the blood brain barrier in three breast cancer brain metastasis models. Breast Cancer Res Treat 144:93–101
33. Khanfir A, Lahiani F, Bouzguenda R, Ayedi I, Daoud J, Frikha M (2013) Prognostic factors and survival in metastatic breast cancer: a single institution experience. Rep Pract Oncol Radiother 18:127–132

Chapter 4
Clinical Perspectives: Breast Cancer Brain Metastasis

Sharon K. Michelhaugh, Aliccia Bollig-Fischer, and Sandeep Mittal

Abstract The incidence of central nervous system metastasis from primary breast cancer has steadily increased with aggressive chemotherapy resulting in improved long-term survival. Brain metastases are more common among those with more aggressive histological subtypes of breast cancer such as triple negative and HER2-positive subtypes. Effectiveness of pharmacological treatment for brain metastases is hindered by the blood–brain barrier. As such, current standard-of-care treatment modalities for CNS metastases include microsurgical resection, whole-brain radiation therapy, and stereotactic radiosurgery, either alone or in combination. Despite providing good local control, involvement of the CNS remains a devastating complication of breast cancer significantly limiting patient survival and quality of life. Leptomeningeal disease is a particularly devastating neurological complication of breast cancer and has limited treatment options. Overall prognosis of breast cancer brain metastasis remains dismal with 1- and 2-year survival rates of 20 % and 2 %, respectively. Clearly, there is a dire need to identify biomarkers permitting earlier and accurate diagnosis of CNS metastases, development of prevention strategies in high-risk individuals, and establishing more effective treatment options such as targeted systemic and intrathecal therapies.

S.K. Michelhaugh
Department of Neurosurgery, Karmanos Cancer Institute, Wayne State University, Detroit, MI, USA

A. Bollig-Fischer
Department of Oncology, Karmanos Cancer Institute, Wayne State University, Detroit, MI, USA

S. Mittal (✉)
Department of Neurosurgery, Karmanos Cancer Institute, Wayne State University, Detroit, MI, USA

Department of Oncology, Karmanos Cancer Institute, Wayne State University, Detroit, MI, USA
e-mail: smittal@med.wayne.edu

© Springer International Publishing Switzerland 2014
S. Sethi, *miRNAs and Target Genes in Breast Cancer Metastasis*,
SpringerBriefs in Cancer Research, DOI 10.1007/978-3-319-08162-5_4

Keywords Brain metastases • Blood–brain barrier • Central nervous system • Treatment • Surgery • Radiosurgery • Whole-brain radiation therapy • Leptomeningeal disease • Clinical trials

Introduction

Central nervous system (CNS) metastases (encompassing the brain, spinal cord, leptomeninges, and retina) represent a devastating aspect of the natural history of a variety of solid malignancies. Over the last two decades, as treatment strategies to control the primary disease became more effective and advanced imaging techniques were further refined and permitted earlier and more accurate diagnosis, development of CNS metastases has become a dreaded neurological complication of patients with systemic cancer. In this chapter, we will discuss the clinical aspects relevant in the management of breast cancer-associated brain metastases and provide an overview of critical elements that may lead to improved future therapies.

Incidence of Breast Cancer Subtypes and Brain Metastases

There are over 200,000 newly diagnosed cases of brain metastases annually in the United States, a tenfold greater incidence than primary brain cancer [1, 2]. Current estimates indicate that up to 30 % of patients with breast cancer will develop a metastatic brain tumor (MBT) in the course of their disease [3, 4]. Of all brain metastases, 20–30 % arise from primary breast cancer, making it the second most common source of MBTs behind lung cancer [3, 5–8]. The frequency at which breast cancer patients develop brain metastases differs according to the histology or molecular subtypes of the primary disease. The luminal A (ER-positive) subtype accounts for over 50 % of primary breast cancers, while the incidence of brain metastases generated from luminal A primary cancers is only 10 %. Conversely, both the HER2-overexpressed and triple negative breast cancer (TNBC) subtypes have disproportionately higher incidences of brain metastases (see Fig. 4.1) [4, 8–13]. In addition to higher incidence, HER2-positive and TNBC subtypes also have a shorter time to onset of MBTs [14]. Improved targeted treatments, such as HER2-targeted trastuzumab, are allowing patients to live longer with stable or progression-free primary cancer, but leading to increased incidence of MBTs [5, 15–17]. In one case report, despite achieving a pathologic complete response for the primary HER2-positive cancer after a treatment regimen including trastuzumab (normally associated with a very favorable prognosis), the patient developed multiple symptomatic MBTs within 5 months [18]. Despite the difference in incidence rates for the individual molecular subtypes, recent findings suggest that there is no difference in patient survival once diagnosed with metastatic CNS disease, which is a dismal 7–9 months [10, 19].

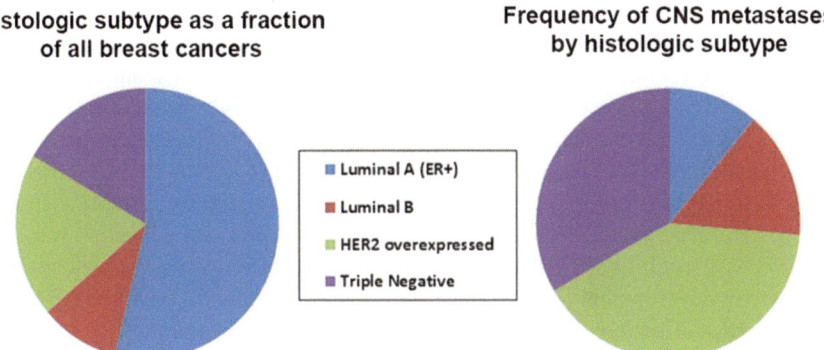

Fig. 4.1 Frequency of occurrence of metastatic brain tumors according to breast cancer histological subtype

Clinical Presentation and Diagnosis

Despite what is known about the breast cancer molecular subtypes and their relative propensities to metastasize to the brain, there are currently no clinical tests or biomarker assays to predict which patients are at risk of developing MBTs. Metastatic lesions are typically discovered when a patient presents with neurological symptoms, most commonly related to increased intracranial pressure (headache, nausea, vomiting), seizures, and/or focal neurological deficits such as weakness on one side of the body or gait abnormalities [20]. These symptoms warrant contrast-enhanced magnetic resonance imaging (MRI) studies, which is the preferred imaging modality when compared to computed tomography (CT) because of its superior soft tissue resolution [21]. Early detection (presymptomatic) of MBTs is possible with MRI and in a very small study ($n = 22$) led to increased patient survival [22], but the healthcare system cannot bear the burden to screen all breast cancer patients with high-resolution MRI scans. Accurate diagnosis of metastatic brain disease is not always possible by MRI, as other conditions such as primary brain tumors (e.g., high-grade gliomas), hematomas, ischemia, and onset of ischemic stroke may have similar imaging characteristics [21]. Final diagnosis of MBT may require histopathologic analyses of biopsied or surgically resected tissue. Molecular imaging using positron emission tomography (PET) is often a useful adjunct in the management of patients with MBTs and may one day provide more accurate diagnosis and improve overall prognosis in patients with MBTs [23–25].

Blood–Brain Barrier

The current standard-of-care treatment for MBTs is limited to surgical resection and radiotherapy because there are no FDA-approved agents for the treatment of breast cancer brain metastases. The chemotherapeutic agents used to treat the primary breast cancer are not effective, partly attributed to their limited ability to penetrate the blood–brain barrier (BBB; [26–28]). While the primary function of the BBB is to regulate passage of proteins, nutrients, electrolytes, and neurotoxins into the CNS, it also precludes entry of a number of chemotherapeutic and targeted agents [10]. Promising preliminary studies indicate that molecularly targeted therapies (small molecule kinase inhibitors against activated oncogenes) can cross the BBB [29, 30] and correlate with modest improved outcomes [31–34]. Selected independent case studies report marked tumor regression and sustained response [35–37]. Although newer therapies that may allow current drugs to penetrate the BBB are in development [38–41], recent evidence also suggests that the BBB may not be intact in patients with primary or metastatic brain tumors [42]. In fact, it seems likely that BBB disruption is part of the metastatic process leading to the development of the brain lesions [43–46].

Molecular Aberrations of Metastatic Brain Tumors

Over 100 years ago, Stephen Paget first proposed that the development of metastases from systemic cancer was not random and that the formation of metastatic tumors depends on interactions between the microenvironment of the metastatic site and the metastatic cancer cells [47, 48]. In order to metastasize, the primary cancer must generate cells that are able to invade or migrate into the bloodstream, and also out of the bloodstream, and survive and colonize in the new host tissue [44–46]. Circulating tumor cells (CTCs) can be detected in the blood of carcinoma patients (including those who do not develop metastases) [49], and despite ~80 % of CTCs undergoing extravasation, fewer than 3 % survive to form micrometastases. Ultimately less than 0.1 % of CTCs will persist through the metastatic colonization process [50]. It also follows that these CTCs would express a phenotype resistant to the treatment given for the primary cancer, contributing to the increased incidence of MBT as described above [5, 15–17] and to the fact that metastases from HER2-positive breast cancers are often HER2-negative [51]. Although large genomic studies of breast cancer-derived MBTs to identify specific genes permissive for colonization in the brain are relatively lacking in the biomedical literature, there are several small studies of gene-coding mutations, somatic copy number variations, and whole-genome expression studies comparing primary breast cancer and brain metastases that seek to identify gene associated specifically with brain metastases (reviewed in [52]). The majority of these studies focused on HER2-positive breast cancer, based on the disproportionate incidence of MBT, while few studies were dedicated to the study of TNBC, despite the equally disproportionate incidence. One example of a neural-related gene that may contribute to metastatic cells ability to colonize in the brain microenvironment is ST6GALNAC5 that encodes a normally brain-specific sialyltransferase that can regulate cell membrane-based cell–cell interactions [53]. Other neuronal proteins identified in MBTs that may play a role in permitting growth in the brain microenvironment are nestin [54], vimentin

[55], and the Wnt signaling pathway [56], all of which play crucial roles in normal development and functioning of neurons and other CNS cell types [57–59].

Current Treatments of CNS Metastases

While there are no specific guidelines for management of CNS metastases from breast cancer, current treatment algorithms are derived from studies which included patients with a variety of solid tumors. Current recommendations for patients with controlled systemic breast cancer include combinations of microsurgical resection, whole-brain radiation therapy (WBRT), and/or stereotactic radiosurgery (SRS), depending on the number of MBTs, the patient's overall performance status, and the extent of systemic disease.

Symptomatic Management

CNS tumors including MBTs can lead to significant neurological deficits due to extensive peritumoral vasogenic cerebral edema. Appropriate use of high-dose corticosteroids can effectively reduce peritumoral edema and associated mass effect often leading to improved neurological symptoms. Moreover, patients with tumor-related seizures need to be managed with judicious use of anticonvulsants to prevent recurrent seizures. Symptomatic management, while not necessarily altering overall prognosis, does represent a critical element of patient care leading to improved quality of life [60].

Limited Disease

For patients with less than 3 metastatic brain lesions, the current recommendation is surgical resection (of the larger, symptomatic tumor) followed by SRS to the resection cavity and other smaller lesions (Fig. 4.2). SRS alone using Gamma Knife, Cyberknife, or linear accelerator-based systems can also provide excellent local tumor control [61]. Treatment response to SRS varies with the molecular subtype of the primary breast cancer. HER2-positive breast cancer seems to have the best response, with TNBC having the worst [62].

Multiple Brain Metastases

For patients with more than 3 lesions, WBRT is generally indicated. However, even in patients with multiple (≥4) lesions and good performance status with limited

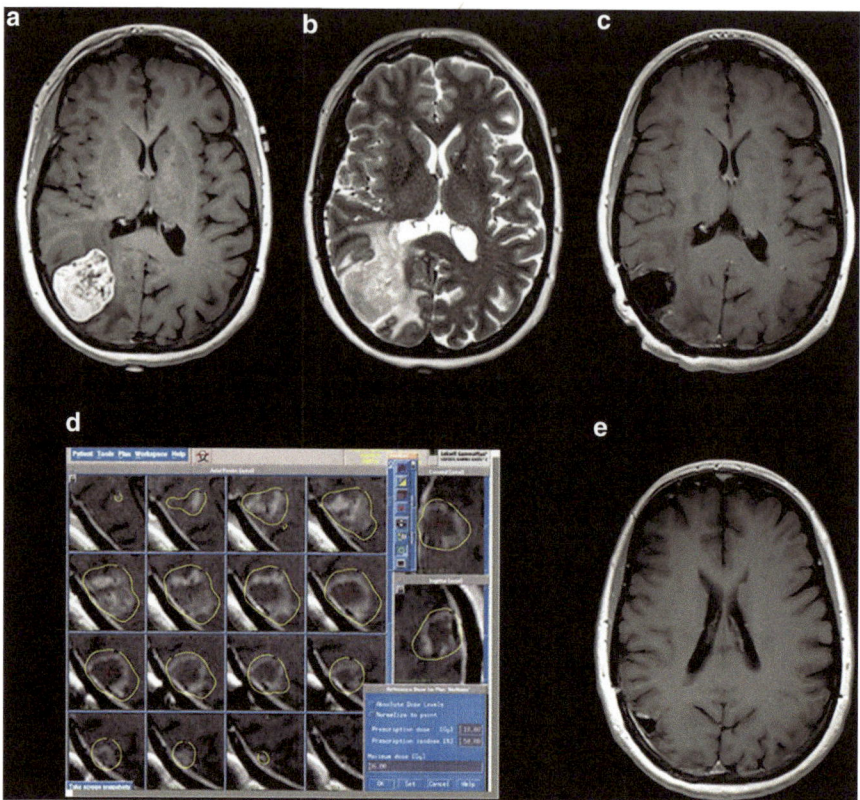

Fig. 4.2 Patient with breast cancer presenting with headaches and new-onset partial seizures diagnosed with a histologically confirmed solitary metastatic brain tumor. Post-contrast T1-weighted MRI (**a**) showing a large right parietal enhancing mass with significant peritumoral vasogenic edema on T2-weighted image (**b**). The patient underwent gross total microsurgical resection of the tumor (**c**). This was followed by Gamma Knife stereotactic radiosurgery to the resection bed 2 weeks later (**d**). Follow-up MRI brain 3 months later showed significant reduction in surgical cavity with no evidence of residual or recurrent tumor (**e**)

extracranial disease, resection of a large, dominant, or symptomatic tumor is occasionally recommended to maintain or improve quality of life followed by WBRT. Patients with active, uncontrolled primary disease and extensive CNS involvement are not typically candidates for surgical resection (Fig. 4.3) [63]. An extensive review of the literature found that adding chemotherapy or radiosensitizers in conjunction with WBRT did not provide any additional benefit to patients [64]. Regardless of these interventions, patients receiving radiation alone have a median survival of 4–6 months, with surgical resection extending median survival to 9–10 months [65, 66].

Fig. 4.3 MRI scan of a patient with innumerable brain metastases from primary breast cancer. There are presently no effective treatment options for this condition

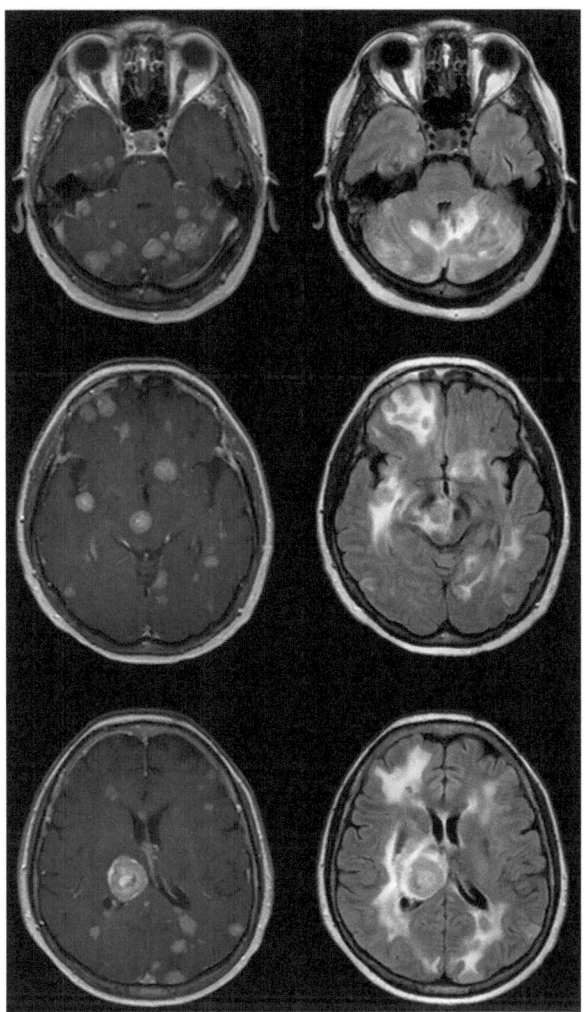

Leptomeningeal Disease

Another CNS complication resulting from metastatic breast cancer is leptomeningeal disease (LMD). As is true with MBTs, the incidence of LMD is increasing as patients live longer with improved systemic therapies resulting in stable primary disease [67]. In the context of metastatic cancer including breast, the neurological symptoms of LMD may present relatively acutely and be highly debilitating. The signs and symptoms vary with the anatomical CNS involvement. Headaches, nausea, seizures, and hydrocephalus are indicative of cerebral involvement. If cranial nerves are involved, patients may experience double vision, loss of visual acuity, hearing loss, or facial numbness. Extremity weakness, tingling

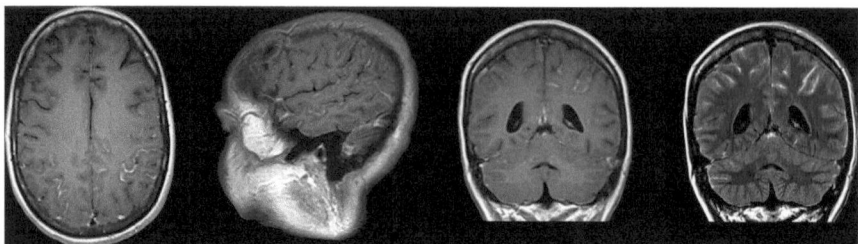

Fig. 4.4 MRI scan of a patient with diffuse leptomeningeal disease from primary breast cancer. There are presently no effective treatment options for this condition

sensations, and/or pain suggest spinal cord involvement. Diagnosis of LMD typically involves contrast-enhanced MRI (Fig. 4.4), as well as demonstration of malignant cells in the cerebrospinal fluid [68, 69]. Treatment may include traditional chemotherapies such as intrathecal or high-dose methotrexate, capecitabine, or platinum-based drugs, although combination therapies may be more effective [67, 70]. Molecular profiling of the malignant cells in the cerebrospinal fluid may yield a better understanding of a metastatic cancer cell subtype that contributes specifically to the development of LMD [71].

Current Clinical Trials

Given the scale of the clinical problem and complexity of MBTs, it is not surprising to find that a search of ClinicalTrials.gov reveals 74 studies that are focused on or include breast cancer brain metastases (as of July 14, 2014). Despite this reasonable number of studies, only five have any available results (Table 4.1), and in the past 5 years, nine studies have been subject to early termination due to inadequate patient enrollment. Most of the studies focus on HER2-positive metastatic breast cancer, with only two studies specific to TNBC. There are also only two studies that include leptomeningeal metastases, whereas most studies list LMD as an exclusion criteria. Most studies are focused on the use of drugs approved for other cancer types or novel combinations of previously approved drugs such as inhibitors targeting HER2 (trastuzumab) or less specific drugs such as lapatinib which target both HER2 and EGFR, antiangiogenic drugs such as bevacizumab, or histone deacetylase inhibitors such as vorinostat, or platinum-based chemotherapy drugs such as carboplatin, or the DNA replication inhibitor irinotecan. Only a few studies are seeking to identify better diagnostic imaging methods or biomarkers with an eye toward earlier detection of MBTs or predicting which patient populations are most at risk of developing MBTs. Until clinical research can focus on better diagnostics and new treatment paradigms that account for the unique biology of brain metastases, the repurposing of other drugs is not likely to provide any improvement in patient outcomes.

Table 4.1 Active studies including breast cancer brain metastasis registered in ClinicalTrials.gov. Only five studies have results available (verified on July 14, 2014)

NCT Number	Title	Recruitment	Study Results
NCT01004172	Carboplatin and Bevacizumab for Progressive Breast Cancer Brain Metastases	Active, not recruiting	No results available
NCT01173497	A Study Evaluating INIPARIB in Combination With Chemotherapy to Treat Triple Negative Breast Cancer Brain Metastasis	Active, not recruiting	No results available
NCT00263588	Lapatinib for Brain Metastases In ErbB2-Positive Breast Cancer	Active, not recruiting	No results available
NCT01077648	Brain Metastasis in Breast Cancer Patients	Completed	No results available
NCT01782274	Proteome-based Immunotherapy of Brain Metastases From Breast Cancer	Active, not recruiting	No results available
NCT01441596	Lux-Breast 3; Afatinib Alone or in Combination With Vinorelbine in Patients With Human Epidermal Growth Factor Receptor 2 (HER2) Positive Breast Cancer Suffering From Brain Metastases	Active, not recruiting	No results available
NCT00617539	Irinotecan and Temozolomide in Treating Patients With Breast Cancer Who Have Received Previous Treatment for Brain Metastases	Active, not recruiting	No results available
NCT00820222	Lapatinib Plus Capecitabine Versus Trastuzumab Plus Capecitabine in ErbB2 (HER2) Positive Metastatic Breast Cancer	Active, not recruiting	Results available
NCT00490139	ALTTO (Adjuvant Lapatinib And/Or Trastuzumab Treatment Optimisation) Study; BIG 2-06/N063D	Active, not recruiting	No results available
NCT00073528	Study Comparing GW572016 And Letrozole Versus Letrozole In Subjects With Advanced Or Metastatic Breast Cancer	Active, not recruiting	Results available
NCT00838929	Study of the Combination of Vorinostat and Radiation Therapy for the Treatment of Patients With Brain Metastases	Active, not recruiting	Results available
NCT00303992	Trastuzumab and Irinotecan in Treating Patients With HER2/Neu Positive Metastatic Breast Cancer	Active, not recruiting	No results available
NCT00470847	Lapatinib in Combination With Radiation Therapy in Patients With Brain Metastases From HER2-Positive Breast Cancer	Completed	Results available
NCT01064349	Breast Cancer With Over-expression of erbB2-BRAINSTORM	Completed	No results available
NCT00098605	Lapatinib in Treating Brain Metastases in Patients With Stage IV Breast Cancer and Brain Metastases	Completed	No results available
NCT00083304	Whole Brain Radiation Therapy With Oxygen, With or Without RSR13, in Women With Brain Metastases From Breast Cancer	Completed	No results available
NCT00614978	Lapatinib and Temozolomide for the Treatment of Progressive Brain Disease in HER-2 Positive Breast Cancer	Completed	No results available
NCT00071383	Analysis of Brain Metastasis in Patients With Breast Cancer, With and Without Over-Expression of HER-2	Completed	No results available
NCT01281696	Bevacizumab With Etoposide and Cisplatin in Breast Cancer Patients With Brain and/or Leptomeningeal Metastasis	Completed	No results available
NCT00639366	Radiation Therapy to the Head in Preventing Brain Metastases in Women Receiving Trastuzumab and Chemotherapy for Metastatic or Locally Advanced Breast Cancer	Completed	No results available
NCT00967031	Lapatinib Ditosylate and Capecitabine in Treating Patients With Stage IV Breast Cancer and Brain Metastases	Completed	No results available
NCT00977379	XERAD: A Study of Xeloda (Capecitabine) Plus Radiotherapy in Patients With Breast Cancer With Newly Diagnosed Brain Metastases	Completed	No results available
NCT00587964	Phase II Trial of Stereotactic Radiosurgery Boost Following Surgical Resection for Brain Metastases	Completed	No results available
NCT00795678	Chemotherapeutic Agents in Brain/Breast	Completed	No results available
NCT01363986	A Study of Herceptin (Trastuzumab) in Combination With Whole Brain Radiotherapy in Patients With HER-2 Positive Breast Cancer	Completed	No results available
NCT00831545	Study to Evaluate the Efficacy and Safety of Temozolomide in Subjects With Brain Metastases of Either Malignant Melanoma, Breast, or Non-Small Cell Lung Cancer (P02064)	Completed	No results available
NCT00916877	Prophylactic Cranial Irradiation in Patients With HER-2-Positive Metastatic Breast Cancer	Completed	No results available
NCT00450866	Epothilone B in Treating Patients With CNS Metastases From Breast Cancer	Completed	Results available
NCT00184275	Characterization of Brain Metastases	Completed	No results available
NCT00539383	A Phase 1, Open-Label, Dose Escalation Study of ANG1005 in Patients With Advanced Solid Tumors and Metastatic Brain Cancer	Completed	No results available

(continued)

Table. 4.1 (continued)

NCT01077726	A Study of Xeloda (Capecitabine) in Breast Cancer Patients With Central Nervous System (CNS) Progression	Completed	No results available
NCT00071357	Use of Dynamic Contrast-Enhanced Magnetic Resonance Imaging to Assess Tumor-Associated Vasculature in Patients With Metastatic Breast Cancer	Completed	No results available
NCT01290354	Exploratory Lapatinib (Positron Emission Tomography) PET Study in Subjects With Breast Cancer	Completed	No results available
NCT01332981	An Observational Follow-up Study of 1st-Line Treatment With Herceptin (Trastuzumab) in Patients With Metastatic Breast Cancer (Post-HERMINE)	Completed	No results available
NCT02048059	ANG1005 in HER2+ Breast Cancer Patients With Progressive/Recurrent Brain Metastases	Recruiting	No results available
NCT02000882	Capecitabine + BKM120 TNBC BC Brain Met	Recruiting	No results available
NCT01924351	HER2-positive Breast Cancer With Brain Metastasis	Not yet recruiting	No results available
NCT02000739	Genetically-informed Therapies for Patients With Metastatic Cancer	Withdrawn	No results available
NCT02038218	Study of 4-Demethyl-4-cholesteryloxycarbonylpenclome in Patients With Brain Tumors	Recruiting	No results available
NCT02005614	A Pilot/Phase II Study of Gamma Knife Radiosurgery for Brain Metastases Using 3Tesla MRI and Rational Dose Selection	Recruiting	No results available
NCT01305941	A Study Of Everolimus, Trastuzumab And Vinorelbine In HER2-Positive Breast Cancer Brain Metastases	Recruiting	No results available
NCT01724606	Whole Brain Radiotherapy (WBRT) With Sorafenib for Breast Cancer Brain Metastases (BCBM)	Recruiting	No results available
NCT01480583	GRN1005 Alone or in Combination With Trastuzumab in Breast Cancer Patients With Brain Metastases	Recruiting	No results available
NCT01942980	Evaluation of the Efficacy of Hippocampal Avoidance on the Cognitive Toxicity of Whole-Brain Radiation Therapy After Surgical Resection of Single Brain Metastasis of Breast Cancer	Recruiting	No results available
NCT01386580	An Open-label, Phase I/IIa, Dose Escalating Study of 2B3-101 in Patients With Solid Tumors and Brain Metastases or Recurrent Malignant Glioma.	Recruiting	No results available
NCT01913067	Evaluation of Cabazitaxel in Patients With Brain Metastasis Secondary to Breast Cancer and NSCLC	Recruiting	No results available
NCT01218529	Lapatinib and WBRT for Patients With Brain Metastases From Lung or Breast Tumors	Recruiting	No results available
NCT01985971	F18 EF5 PET/CT Imaging in Patients With Brain Metastases From Breast Cancer	Recruiting	No results available
NCT01921335	Phase I Dose-escalation Trial of ARRY-380 in Combination With Trastuzumab in Participants With Brain Metastases From HER2+ Breast Cancer	Recruiting	No results available
NCT00875355	Radiation Therapy With or Without Temozolomide in Treating Women With Brain Metastases and Breast Cancer	Recruiting	No results available
NCT01622868	Whole-Brain Radiation Therapy With or Without Lapatinib Ditosylate in Treating Patients With Brain Metastasis From HER2-Positive Breast Cancer	Recruiting	No results available
NCT01494662	HKI-272 for HER2-Positive Breast Cancer and Brain Metastases	Recruiting	No results available
NCT00377156	Stereotactic Radiation Therapy With or Without Whole-Brain Radiation Therapy in Treating Patients With Brain Metastases	Recruiting	No results available
NCT01132664	Safety and Efficacy of BKM120 in Combination With Trastuzumab in Patients With Relapsing HER2 Overexpressing Breast Cancer Who Have Previously Failed Trastuzumab	Recruiting	No results available
NCT00398437	Magnetic Resonance Imaging for the Early Detection of CNS Metastases in Women With Stage IV Breast Cancer	Recruiting	No results available
NCT01621906	18F-FLT-PET Imaging of the Brain in Patients With Metastatic Breast Cancer to the Brain Treated With Whole Brain Radiation Therapy With or Without Sorafenib: Comparison With MR Imaging of the Brain	Recruiting	No results available
NCT01332630	TPI 287 in Breast Cancer Metastatic to the Brain	Recruiting	No results available
NCT00570908	Brain Mets - Capecitabine and WBRT	Terminated	No results available

(continued)

Table. 4.1 (continued)

NCT01939483	A Pilot Study of Irinotecan in Patients With Breast Cancer and CNS Metastases	Recruiting	No results available
NCT00637637	External-Beam Radiation Therapy With or Without Indinavir and Ritonavir in Treating Patients With Brain Metastases	Recruiting	No results available
NCT01706432	Hypofractionated Image Guided Radiation Therapy in Treating Patients With Stage IV Breast Cancer	Active, not recruiting	No results available
NCT01690702	Study of Nab-Paclitaxel in High Risk Early Breast Cancer	Recruiting	No results available
NCT00992602	High-dose Methotrexate and Liposomal Cytarabine in Treating Patients With CNS Metastases From Breast Cancer	Recruiting	No results available
NCT01783756	Phase 1b/2 Trial Using Lapatinib, Everolimus and Capecitabine for Treatment of HER-2 Positive Breast Cancer With CNS Metastasis	Recruiting	No results available
NCT01818713	Clinical and Pharmacological Study With 2B3-101 in Patients With Breast Cancer and Leptomeningeal Metastases	Recruiting	No results available
NCT01934894	Cabazitaxel Plus Lapatinib as Therapy for HER2-Positive Metastatic Breast Cancer Patients With Intracranial Metastases	Recruiting	No results available
NCT00188864	Dexamethasone for Palliation - Brain Metastases	Recruiting	No results available
NCT01806675	18F FPPRGD2 Positron Emission Tomography/Computed Tomography in Predicting Early Response in Patients With Cancer Receiving Anti-Angiogenesis Therapy	Recruiting	No results available
NCT01710605	Medico-economic Interest of Taking Into Account Circulating Tumor Cells (CTC) to Determine the Kind of First Line Treatment for Metastatic, Hormone-receptors Positive, Breast Cancers	Recruiting	No results available
NCT01414933	High Throughput Technologies to Drive Breast Cancer Patients to Specific Phase I/II Trials of Targeted Agents	Completed	No results available
NCT02154529	Study of the Combination of KD019 and Trastuzumab in Subjects With HER2-Postive Metastatic Breast Cancer	Recruiting	No results available
NCT02133677	A Phase II Multi-center Pilot Study of Concurrent Temozolomide and Whole Brain Irradiation in Lung Cancer and Breast Cancer Patients With Brain Metastases	Active, not recruiting	No results available
NCT02185352	Bevacizumab, Etoposide and Cisplatin Followed by Whole Brain Radio-therapy in Breast Cancer With Brain Metastases	Active, not recruiting	No results available
NCT02166658	A Study of Cabazitaxel for Patients With Breast or Lung Cancer and Recurrent or Progressive Brain Metastases - Cabazitaxel for Brain Metastases (CaBaMet)	Active, not recruiting	No results available

MicroRNAs as Targets for Future Diagnostic, Prognostic, and Therapeutic Strategies

There is a clear and dire need for better diagnostics and treatments for breast cancer-derived MBTs. MicroRNAs might prove to be a useful avenue to pursue in this regard. As an example, miR-21 has been shown to have higher expression levels in invasive breast cancer cells vs. noninvasive cells [72] and is associated with a shorter disease-free interval before relapse [73]. Intriguingly, miR-21 (and also miR-10b) expression is also increased in the cerebrospinal fluid of glioblastoma patients (a high-grade primary brain tumor known to be highly invasive), as well as patients with MBTs, while miR-200 family members are elevated in the cerebrospinal fluid of patients with MBTs only [74] which may allow for a relatively simple laboratory test to aid in the differential diagnosis of patients with enhancing masses on MRI. Although not demonstrated specifically in brain metastases, miR-301a was shown by Ma et al. to maintain activated Wnt signaling, promoting breast cancer invasion and metastasis [75], which is congruent with an identified role of Wnt signaling in metastases from TNBC [56]. The specific roles of microRNAs in the development of metastases, biomarkers for diagnosis and prognosis, and as therapeutic targets requires further investigation.

Conclusion

In conclusion, MBTs originating from breast cancer remain a challenging clinical problem, with a paucity of efficacious treatment options. Going forward, research studies and clinical trials that advance neuroimaging technology and better define the genomics and underlying biology of MBTs should advance the cause of improved diagnostics and targeted treatments and ultimately achieve better patient outcomes.

References

1. Gavrilovic IT, Posner JB (2005) Brain metastases: epidemiology and pathophysiology. J Neurooncol 75:5–14
2. Patchell RA (2003) The management of brain metastases. Cancer Treat Rev 29:533–540
3. Nathoo N, Chahlavi A, Barnett GH, Toms SA (2005) Pathobiology of brain metastases. J Clin Pathol 58:237–242
4. Tsukada Y, Fouad A, Pickren JW, Lane WW (1983) Central nervous system metastasis from breast carcinoma. Autopsy study. Cancer 52:2349–2354
5. Norden AD, Wen PY, Kesari S (2005) Brain metastases. Curr Opin Neurol 18:654–661
6. Wen PY, Loeffler JS (2000) Brain metastases. Curr Treat Opt Oncol 1:447–458
7. Niwinska A, Murawska M, Pogoda K (2010) Breast cancer subtypes and response to systemic treatment after whole-brain radiotherapy in patients with brain metastases. Cancer 116:4238–4247
8. Chang EL, Lo S (2003) Diagnosis and management of central nervous system metastases from breast cancer. Oncologist 8:398–410
9. Langer CJ, Mehta MP (2005) Current management of brain metastases, with a focus on systemic options. J Clin Oncol 23:6207–6219
10. Dawood S, Gonzalez-Angulo AM (2013) Progress in the biological understanding and management of breast cancer-associated central nervous system metastases. Oncologist 18:675–684. doi:10.1634/theoncologist.2012-0438
11. Pogoda K, Niwinska A, Murawska M, Pienkowski T (2013) Analysis of pattern, time and risk factors influencing recurrence in triple-negative breast cancer patients. Med Oncol 30:388
12. Steward L, Conant L, Gao F, Margenthaler JA (2014) Predictive factors and patterns of recurrence in patients with triple negative breast cancer. Ann Surg Oncol 21:21
13. Harrell JC, Prat A, Parker JS, Fan C, He X, Carey L et al (2012) Genomic analysis identifies unique signatures predictive of brain, lung, and liver relapse. Breast Cancer Res Treat 132:523–535
14. Ahn HK, Park YH, Lee SJ, Park S, Maeng CH, Park W et al (2013) Clinical implication of Time To Brain Metastasis (TTBM) according to breast cancer subtypes. Springerplus 2:136

15. Burstein HJ, Lieberman G, Slamon DJ, Winer EP, Klein P (2005) Isolated central nervous system metastases in patients with HER2-overexpressing advanced breast cancer treated with first-line trastuzumab-based therapy. Ann Oncol 16:1772–1777
16. Kaplan MA, Ertugrul H, Firat U, Kucukoner M, Inal A, Urakci Z et al (2014) Brain metastases in HER2-positive metastatic breast cancer patients who received chemotherapy with or without trastuzumab. Breast Cancer. doi:10.1007/s12282-013-0513-z
17. Bendell JC, Domchek SM, Burstein HJ, Harris L, Younger J, Kuter I et al (2003) Central nervous system metastases in women who receive trastuzumab-based therapy for metastatic breast carcinoma. Cancer 97:2972–2977
18. Shimada K, Ishikawa T, Yoneyama S, Kita K, Narui K, Sugae S et al (2013) Early-onset brain metastases in a breast cancer patient after pathological complete response to neoadjuvant chemotherapy. Anticancer Res 33:5119–5121
19. Quigley MR, Fukui O, Chew B, Bhatia S, Karlovits S (2013) The shifting landscape of metastatic breast cancer to the CNS. Neurosurg Rev 36:377–382
20. Andrade F, Aguiar PH, Fontes RB, Nakagawa E, Teixeira JA, Miura FK et al (2004) Clinical presentation, treatment and outcome of patients with cerebral metastases: the University of Sao Paulo series. Arq Neuropsiquiatr 62:808–814
21. Barajas RF Jr, Cha S (2012) Imaging diagnosis of brain metastasis. Prog Neurol Surg 25:55–73
22. Mystakidou K, Kouloulias V, Tsilika E, Boviatsis E, Kouvaris J, Matsopoulos G et al (2004) Is early recognition of radiologically silent brain metastasis from breast cancer beneficial? A retrospective study of 22 cases. Breast Cancer 11:276–281
23. Juhasz C, Dwivedi S, Kamson DO, Michelhaugh SK, Mittal S (2014) Comparison of amino acid positron emission tomographic radiotracers for molecular imaging of primary and metastatic brain tumors. Mol Imaging 13:1–16
24. Kamson DO, Mittal S, Buth A, Muzik O, Kupsky WJ, Robinette NL et al (2013) Differentiation of glioblastomas from metastatic brain tumors by tryptophan uptake and kinetic analysis: a positron emission tomographic study with magnetic resonance imaging comparison. Mol Imaging 12:327–337
25. Juhasz C, Nahleh Z, Zitron I, Chugani DC, Janabi MZ, Bandyopadhyay S et al (2012) Tryptophan metabolism in breast cancers: molecular imaging and immunohistochemistry studies. Nucl Med Biol 39:926–932
26. Greig NH, Soncrant TT, Shetty HU, Momma S, Smith QR, Rapoport SI (1990) Brain uptake and anticancer activities of vincristine and vinblastine are restricted by their low cerebrovascular permeability and binding to plasma constituents in rat. Cancer Chemother Pharmacol 26:263–268
27. Genka S, Deutsch J, Stahle PL, Shetty UH, John V, Robinson C et al (1990) Brain and plasma pharmacokinetics and anticancer activities of cyclophosphamide and phosphoramide mustard in the rat. Cancer Chemother Pharmacol 27:1–7
28. Mehta AI, Brufsky AM, Sampson JH (2013) Therapeutic approaches for HER2-positive brain metastases: circumventing the blood-brain barrier. Cancer Treat Rev 39:261–269
29. Heimberger AB, Learn CA, Archer GE, McLendon RE, Chewning TA, Tuck FL et al (2002) Brain tumors in mice are susceptible to blockade of epidermal growth factor receptor (EGFR) with the oral, specific, EGFR-tyrosine kinase inhibitor ZD1839 (iressa). Clin Cancer Res 8:3496–3502
30. Gluck S, Castrellon A (2009) Lapatinib plus capecitabine resolved human epidermal growth factor receptor 2-positive brain metastases. Am J Ther 16:585–590
31. Geyer CE, Forster J, Lindquist D, Chan S, Romieu CG, Pienkowski T et al (2006) Lapatinib plus capecitabine for HER2-positive advanced breast cancer. N Engl J Med 355:2733–2743
32. Lin NU, Dieras V, Paul D, Lossignol D, Christodoulou C, Stemmler HJ et al (2009) Multicenter phase II study of lapatinib in patients with brain metastases from HER2-positive breast cancer. Clin Cancer Res 15:1452–1459
33. Narayana A, Mathew M, Tam M, Kannan R, Madden KM, Golfinos JG et al (2013) Vemurafenib and radiation therapy in melanoma brain metastases. J Neurooncol 113:411–416

34. Mok T, Yang JJ, Lam KC (2013) Treating patients with EGFR-sensitizing mutations: first line or second line–is there a difference? J Clin Oncol 31:1081–1088

35. Abboud M, Saghir NS, Salame J, Geara FB (2010) Complete response of brain metastases from breast cancer overexpressing Her-2/neu to radiation and concurrent Lapatinib and Capecitabine. Breast J 16:644–646

36. Slamon DJ, Leyland-Jones B, Shak S, Fuchs H, Paton V, Bajamonde A et al (2001) Use of chemotherapy plus a monoclonal antibody against HER2 for metastatic breast cancer that overexpresses HER2. N Engl J Med 344:783–792

37. Gounant V, Wislez M, Poulot V, Khalil A, Lavole A, Cadranel J et al (2007) Subsequent brain metastasis responses to epidermal growth factor receptor tyrosine kinase inhibitors in a patient with non-small-cell lung cancer. Lung Cancer 58:425–428

38. Park EJ, Zhang YZ, Vykhodtseva N, McDannold N (2012) Ultrasound-mediated blood-brain/blood-tumor barrier disruption improves outcomes with trastuzumab in a breast cancer brain metastasis model. J Control Release 163:277–284

39. Cote J, Bovenzi V, Savard M, Dubuc C, Fortier A, Neugebauer W et al (2012) Induction of selective blood-tumor barrier permeability and macromolecular transport by a biostable kinin B1 receptor agonist in a glioma rat model. PLoS One 7:e37485

40. Caffo M, Barresi V, Caruso G, Cutugno M, La Fata G, Venza M et al (2013) Innovative therapeutic strategies in the treatment of brain metastases. Int J Mol Sci 14:2135–2174

41. Connell JJ, Chatain G, Cornelissen B, Vallis KA, Hamilton A, Seymour L et al (2013) Selective permeabilization of the blood-brain barrier at sites of metastasis. J Natl Cancer Inst 105:1634–1643

42. Nduom EK, Yang C, Merrill MJ, Zhuang Z, Lonser RR (2013) Characterization of the blood-brain barrier of metastatic and primary malignant neoplasms. J Neurosurg 119:427–433

43. Wilhelm I, Molnar J, Fazakas C, Hasko J, Krizbai IA (2013) Role of the blood-brain barrier in the formation of brain metastases. Int J Mol Sci 14:1383–1411

44. Fidler IJ (2011) The role of the organ microenvironment in brain metastasis. Semin Cancer Biol 21:107–112

45. Valastyan S, Weinberg RA (2011) Tumor metastasis: molecular insights and evolving paradigms. Cell 147:275–292

46. Kang Y, Pantel K (2013) Tumor cell dissemination: emerging biological insights from animal models and cancer patients. Cancer Cell 23:573–581

47. Paget S (1989) The distribution of secondary growths in cancer of the breast. 1889. Cancer Metastasis Rev 8:98–101

48. Ramakrishna R, Rostomily R (2013) Seed, soil, and beyond: The basic biology of brain metastasis. Surg Neurol Int 4:S256–S264

49. Nagrath S, Sequist LV, Maheswaran S, Bell DW, Irimia D, Ulkus L et al (2007) Isolation of rare circulating tumour cells in cancer patients by microchip technology. Nature 450:1235–1239

50. Luzzi KJ, MacDonald IC, Schmidt EE, Kerkvliet N, Morris VL, Chambers AF et al (1998) Multistep nature of metastatic inefficiency: dormancy of solitary cells after successful extravasation and limited survival of early micrometastases. Am J Pathol 153:865–873

51. Lower EE, Glass E, Blau R, Harman S (2009) HER-2/neu expression in primary and metastatic breast cancer. Breast Cancer Res Treat 113:301–306

52. Bollig-Fischer A, Michelhaugh SK, Ali-Fehmi R, Mittal S (2013) The molecular genomics of metastatic brain tumors. OA Mol Oncol 1:6

53. Bos PD, Zhang XH, Nadal C, Shu W, Gomis RR, Nguyen DX et al (2009) Genes that mediate breast cancer metastasis to the brain. Nature 459:1005–1009

54. Sihto H, Lundin J, Lundin M, Lehtimaki T, Ristimaki A, Holli K et al (2011) Breast cancer biological subtypes and protein expression predict for the preferential distant metastasis sites: a nationwide cohort study. Breast Cancer Res 13:R87

55. Brogi E, Murphy CG, Johnson ML, Conlin AK, Hsu M, Patil S et al (2011) Breast carcinoma with brain metastases: clinical analysis and immunoprofile on tissue microarrays. Ann Oncol 22:2597–2603

56. Dey N, Barwick BG, Moreno CS, Ordanic-Kodani M, Chen Z, Oprea-Ilies G et al (2013) Wnt signaling in triple negative breast cancer is associated with metastasis. BMC Cancer 13:537
57. Gilyarov AV (2008) Nestin in central nervous system cells. Neurosci Behav Physiol 38:165–169
58. Lepinoux-Chambaud C, Eyer J (2013) Review on intermediate filaments of the nervous system and their pathological alterations. Histochem Cell Biol 140:13–22
59. Dickins EM, Salinas PC (2013) Wnts in action: from synapse formation to synaptic maintenance. Front Cell Neurosci 7:162
60. Arslan C, Dizdar O, Altundag K (2010) Systemic treatment in breast-cancer patients with brain metastasis. Expert Opin Pharmacother 11:1089–1100
61. Lippitz B, Lindquist C, Paddick I, Peterson D, O'Neill K, Beaney R (2014) Stereotactic radiosurgery in the treatment of brain metastases: the current evidence. Cancer Treat Rev 40:48–59
62. Vern-Gross TZ, Lawrence JA, Case LD, McMullen KP, Bourland JD, Metheny-Barlow LJ et al (2012) Breast cancer subtype affects patterns of failure of brain metastases after treatment with stereotactic radiosurgery. J Neurooncol 110:381–388
63. Gil-Gil MJ, Martinez-Garcia M, Sierra A, Conesa G, Del Barco S, Gonzalez-Jimenez S et al (2014) Breast cancer brain metastases: a review of the literature and a current multidisciplinary management guideline. Clin Transl Oncol 16(5):436–446
64. Tsao MN, Lloyd N, Wong RK, Chow E, Rakovitch E, Laperriere N et al (2012) Whole brain radiotherapy for the treatment of newly diagnosed multiple brain metastases. Cochrane Database Syst Rev 4:CD003869
65. Patchell RA, Tibbs PA, Walsh JW, Dempsey RJ, Maruyama Y, Kryscio RJ et al (1990) A randomized trial of surgery in the treatment of single metastases to the brain. N Engl J Med 322:494–500
66. Vecht CJ, Haaxma-Reiche H, Noordijk EM, Padberg GW, Voormolen JH, Hoekstra FH et al (1993) Treatment of single brain metastasis: radiotherapy alone or combined with neurosurgery? Ann Neurol 33:583–590
67. Jo JC, Kang MJ, Kim JE, Ahn JH, Jung KH, Gong G et al (2013) Clinical features and outcome of leptomeningeal metastasis in patients with breast cancer: a single center experience. Cancer Chemother Pharmacol 72:201–207
68. Leal T, Chang JE, Mehta M, Robins HI (2011) Leptomeningeal metastasis: challenges in diagnosis and treatment. Curr Cancer Ther Rev 7:319–327
69. Scott BJ, Kesari S (2013) Leptomeningeal metastases in breast cancer. Am J Cancer Res 3:117–126
70. Mego M, Sycova-Mila Z, Obertova J, Rajec J, Liskova S, Palacka P et al (2011) Intrathecal administration of trastuzumab with cytarabine and methotrexate in breast cancer patients with leptomeningeal carcinomatosis. Breast 20:478–480
71. Magbanua MJ, Melisko M, Roy R, Sosa EV, Hauranieh L, Kablanian A et al (2013) Molecular profiling of tumor cells in cerebrospinal fluid and matched primary tumors from metastatic breast cancer patients with leptomeningeal carcinomatosis. Cancer Res 73:7134–7143
72. Petrovic N, Mandusic V, Stanojevic B, Lukic S, Todorovic L, Roganovic J et al (2014) The difference in miR-21 expression levels between invasive and non-invasive breast cancers emphasizes its role in breast cancer invasion. Med Oncol 31:867
73. Markou A, Yousef GM, Stathopoulos E, Georgoulias V, Lianidou E (2014) Prognostic significance of metastasis-related microRNAs in early breast cancer patients with a long follow-up. Clin Chem 60:197–205
74. Teplyuk NM, Mollenhauer B, Gabriely G, Giese A, Kim E, Smolsky M et al (2012) MicroRNAs in cerebrospinal fluid identify glioblastoma and metastatic brain cancers and reflect disease activity. Neuro Oncol 14:689–700
75. Ma F, Zhang J, Zhong L, Wang L, Liu Y, Wang Y et al (2014) Upregulated microRNA-301a in breast cancer promotes tumor metastasis by targeting PTEN and activating Wnt/beta-catenin signaling. Gene 535:191–197

Chapter 5
Clinical Perspectives: Breast Cancer Bone Metastasis

Allen Kadado, Anil Sethi, and Rahul Vaidya

Abstract Bone metastasis of breast cancer denotes progression of the disease process resulting in increased morbidity and a decrease in quality of life. Early diagnosis and treatment may help minimize pain and improve function. Metastatic lesions from the breast are commonly osteolytic but may be osteoblastic or mixed in nature and are mediated by the RANKL or endothelin1 pathway. The spine and proximal femur are the commonest sites of involvement. Patients are often in pain which is made worse by pathological fractures. Spinal cord compression may occur in vertebral fractures with the presence of neurological deficits. A complete workup is essential to delineate the metastatic lesion and determine the structural integrity of the pathological site. Bisphosphonates decrease osteoclastic activity and are administered to delay or prevent skeletal events. Indications for surgical management of metastatic bone disease include impending fractures, pathologic fractures of long bones and pelvis, untreatable bone pain, instability, and spinal cord compression. Spinal decompression and stabilization is indicated if the metastatic lesion is accompanied with weakness in the extremities. Impending or complete fractures of the long bones will necessitate internal fixation permitting early mobilization and pain relief. Surgical treatment of bone metastasis has not been reported to improve survival.

Keywords Breast cancer • Bone metastasis • Spinal decompression • Internal fixation

A. Kadado • A. Sethi (✉)
Department of Orthopedic Surgery, Detroit Receiving Hospital, UHC Suite 4G, 4201 St Antoine Blvd, Detroit, MI 48201, USA
e-mail: anilsethi09@gmail.com

R. Vaidya
Department of Orthopedic Surgery, Detroit Receiving Hospital, UHC Suite 4G, 4201 St Antoine Blvd, Detroit, MI 48201, USA

Karmanos Cancer Institute, Detroit, MI, USA

© Springer International Publishing Switzerland 2014
S. Sethi, *miRNAs and Target Genes in Breast Cancer Metastasis*,
SpringerBriefs in Cancer Research, DOI 10.1007/978-3-319-08162-5_5

Introduction

Breast cancer remains the commonest malignancy occurring in women and the second leading cause of death [1]. Improvement in the survival of cancer patients has led to an increased incidence of metastatic bone disease [2]. Between 2001 and 2010 death rates due to cancer declined in women by 1.4 % [3]. Postmortem studies have shown that 73 % of individuals who die with breast cancer have bone metastases, whether or not they were diagnosed antemortem. The average time from diagnosis of breast cancer to diagnosis of bone metastases is 4.3 years [1].

Breast cancer metastases to bone are commonly lytic but may be blastic in nature. The lytic lesions are mediated through the receptor activator of nuclear factor kappa-B ligand (RANKL) pathway activated by tumor produced parathyroid hormone-related protein (PTHrP) [4]. Osteoblastic metastases are produced due to tumor secreted endothelin 1. The use of bisphosphonates in the treatment of cancer has resulted in a decrease in the frequency of skeletal events improving quality of life and clinical outcome. Additionally, the availability of new bone-targeted molecules such as denosumab has added further impetus in this direction [4]. However once bone metastasis occurs its management is a clinical challenge and may include surgical decisions in association with medical treatment with an aim to minimize pain and disability.

The optimal management of skeletal metastases depends on the underlying biology of the disease, location, and extent of bone involvement; the availability of effective systemic therapies and life expectancy of the patient [5]. It is then imperative for the treating physicians to be cognizant of the clinical presentation and treatment options available for better outcomes. The overlying goal of surgical intervention in breast cancer bone metastases is to increase patient survival and quality of life by decreasing pain and restoring function. The purpose of this paper is to discuss the clinical presentation, diagnosis, and available surgical treatment based on the presentation of skeletal involvement.

Clinical Features

The comorbidities associated with the spread of cancer to bone are well understood. These include pain, pathologic fracture, hypercalcemia, and neural compression including spinal cord compression [2, 6].

Pain: Bone metastases frequently result in pain which progressively increases in intensity. Pain is initially localized to the metastatic site but may engulf the extremity following a pathological fracture. Constant pain persistent at night, not alleviated by change of position or lying down, is the hallmark of tumor or infection. Pain is probably the result of tumor-induced osteolysis, tumor production of growth factors and cytokines, direct infiltration of nerves, stimulation of ion channels, and local tissue production of endothelins and nerve growth factors [7].

Pathological Fracture: The vertebral column is commonly affected with meta-static disease which may result in compression fractures causing back pain. In 10 % of these patients the spine may be unstable [7]. These patients often note pain in the back which is worse with activity. Metastatic fractures of long bones usually result in loss of function of the extremity. They are preceded by trabecular disruption and micro fractures which cause complete breaks spontaneously or with minor trauma [6]. When present in the lower extremities they are accompanied with an inability to bear weight on the leg. The extremity may display swelling but this is less marked than following a fracture resulting from trauma.

Spinal cord compression: This is an uncommon complication in metastatic disease of the breast and has a reported incidence of 3–4 % [8, 9]. However, it causes considerable disability and prolonged hospitalization if not recognized early and treated effectively [9]. Patients may develop motor and sensory loss with incontinence warranting emergent surgery. The most common reported symptom of spinal cord compression is motor weakness, followed by pain, sensory distur-bance, and sphincter abnormalities [9]. Presence of spinal instability due to a destructive lesion or following decompression requires stabilization with spinal instrumentation.

Hypercalcemia: In many patients, hypercalcemia is a result of bone destruction. Patients may present with nonspecific symptoms including fatigue, anorexia, and constipation. A rising serum calcium level may lead to deterioration of renal function and mental status.

A high index of suspicion should be maintained to diagnose it early delaying mortality from cardiac arrhythmia and renal failure.

Tumor Characteristics

The process of breast cancer metastasizing to bone is both complex and organized. Cancer cells must first grow within the primary tumor, invade adjacent basement membrane and nearby vasculature, survive in the circulatory system, extravagate into distant tissue, and initiate and maintain an environment for growth [10]. This operation requires the contribution of various genetic mutations, biochemical reactions, and altered signaling pathways. Some overexpressed genes in this pro-gression include CXCR4, FGF-5, IL-11, MMP, ADAMTS1, and proteoglycan-1 [11].

Cooperating with metastasized cancer cells, bone serves a hospitable environ-ment for breast cancer metastases. Bone is a highly vascular organ with greater vascularity in the axial skeleton. It also offers an abundance of chemokines and growth factors which can aid in tumor survival and growth.

Matrix metalloproteinase and the chemokine family are crucial in the metastasis of breast cancer to bone. MMP is involved in the invasion of basement membrane by cooperating with adhesion molecules. MMP is significantly increased in HER2/ neu positive tumors, which may help explain why this phenotype has a poor prognosis [12]. Two chemokines that are involved in the migration of breast cancer to bone are CXCR4 and CCR7. CXCR4 is expressed almost four times as often in breast cancer with bone metastases than breast cancer without bone metastases [13]. CCR7 is expressed exclusively in breast cancer with bone metastases and is never expressed in nonmetastatic breast cancer. CCL2 and CCL5 are chemokines that promote angiogenesis and interactions between cancer cells and bone [14].

Uniquely, breast cancer bone metastases may be purely osteoblastic, purely osteolytic, or offer a mixed picture of osteoblastic and osteolytic lesions. About 48 % are solely osteolytic, 13 % solely osteoblastic, and 38 % mixed [15]. It is understood that the osteoblastic and osteolytic changes are two separate processes that may occur simultaneously. PTHrP plays a crucial role in osteolysis as it helps up regulate RANK-L from osteoblasts. This further increases osteoclastic activity and promotes other tumor genesis processes [16]. Osteoblastic lesions are less understood, but it is agreed that they appear to be stimulated by tumor secreted endothelin-1 [17]. Endothelin-1 is involved in a cascade of events which ultimately leads to increased activity of osteoblasts. Interestingly, PTHrP is also overexpressed in osteoblastic lesions, but it is hypothesized that PTHrP undergoes proteolytic cleavage to mimic the same osteoblastic pathway stimulated by endothelin-1 [18].

The histopathologic grade of primary breast cancer also seems to play a role in its metastatic potential to bone. Breast cancer bone metastases are more common with grade 1 well-differentiated tumors, as compared to the less differentiated phenotypes [19]. It can be postulated that these well-differentiated tumors offer more organized and succinct pathologic pathways as outlined above.

The hormonal receptor status of the primary breast cancer also appears to be important in the metastatic potential to bone. Among the breast cancers that metastasize to bone, 80 % are estrogen receptor positive, 60 % are progesterone receptor positive, and 57 % being both estrogen receptor and progesterone receptor positive [8]. HER2/neu positive also have greater metastatic potential [8].

Diagnostic Work Up

History and physical exam which includes personal history, family history, and a thorough review of systems are imperative and should be performed on any patient with possible metastatic bone disease. Understanding the patient's passage through the disease is crucial, as treatment of metastatic disease is multidisciplinary and requires good teamwork.

Laboratory studies include a complete blood-cell count with differential, ESR, CRP, renal, and liver function tests as also serum calcium estimation. Prostate-specific antigen, carcinoembryonic antigen, serum protein electrophoresis, and immunoelectrophoresis are also helpful especially if the primary lesion is unknown. A urinalysis including test for Bence Jones proteins should also be performed.

Tissue from the pathological fracture site obtained during surgical stabilization may direct towards an unknown primary.

Imaging studies should include radiographs; CT of chest, abdomen, and pelvis; and a whole-body bone scan. MRI is helpful in evaluating spinal lesions and also the soft tissues in the appendicial skeleton. Patients presenting with a known diagnosis of breast cancer with musculoskeletal pain will require imaging studies to assess the metastatic deposit for surgical planning.

Technetium bone scan is useful in determining extent of disease. Skeletal disease is considered widespread if five or more foci are discovered with at least one focus confirmed by radiograph. Minimal skeletal disease is considered when fewer than five foci are involved [8]. Routine bone scan in early stages of breast cancer is not cost-effective [20] since bone relapse occurs only in 8 % of patients presenting with primary breast cancer.

Nonsurgical Treatment

Metastatic breast cancer can be managed nonsurgically with hormonal therapy, chemotherapy, radiation, or a combination of these modalities. Bisphosphonate therapy has become an integral part of treatment in patients with metastatic bone disease. They inhibit bone resorption through inhibition and apoptosis of osteoclasts. In high concentrations, they may also inhibit osteoblasts and tumor cells, including breast cancer cells [21]. Bisphosphonates have been shown to significantly reduce frequency of skeletal-related events, prolong the time to skeletal-related events, and improve bone pain and hypercalcemia [22]. However, bisphosphonates have not been shown to prevent incidence of new bone metastases, or ultimately affect the survival in women with metastatic breast cancer to bone [23]. Despite bisphosphonates lack of effect on mortality, it is still imperative for patients with breast cancer metastatic to bone be considered for bisphosphonate therapy due to their significant improvement of morbidity. Bisphosphonate therapy should be initiated in patients who have known breast cancer with evidence of bony destruction on radiologic imaging [24]. Therapy may also continue beyond 2 years depending on the individual risk of a patient, despite the increased skeletal morbidity associated with prolonged bisphosphonate use [24].

Radiotherapy is utilized in the treatment of bone pain and pathologic fractures associated with bone metastases. Aside from opioids and analgesic medications, radiotherapy is often considered first line in the treatment of bone pain. It has been shown to improve bone pain within weeks [25], and can have a synergistic effect if used alongside bisphosphonate therapy [25]. Radiotherapy can also be used in conjunction with surgery to synergistically improve skeletal morbidity. It also plays a role in the management of spinal cord compression, especially in those who are deemed unfit for surgery [22].

Surgical Management

Most common sites of skeletal involvement are spine and proximal femur, followed by humerus, ribs, and pelvis. Fractures due to bone metastases from breast cancer respond better to surgery than bone metastases of other primary tumors [26]. Indications for surgical management of metastatic bone disease include impending fractures, pathologic fractures of long bones and pelvis, untreatable bone pain, instability, and spinal cord compression. Surgical approaches vary depending upon fracture location, fracture type, type of lesion, and associated symptoms. Femur is the most common long bone to exhibit pathologic fracture, followed by humerus [27]. Surgical options for proximal femur metastases & pathological fractures depend on type & location of lesion. Surgical approaches should be evaluated against current guidelines for fixation. Options include hip prosthetic replacement, intramedullary nailing, and plate & screw fixation. Intramedullary nailing for pathological femoral diaphyseal or metaphyseal fractures have been shown to improve quality of life by offering stability, pain relief, and early post-op mobilization and ambulation [28]. Surgical options for humeral fractures also depend on anatomical location, and include closed reduction & fixation, unreamed intramedullary nailing, plating with screws, and endoprosthetic replacement. Unreamed nailing delivers immediate stability and pain relief with early functional return [29]. Endoprosthetic replacement ensues functionality of the upper limb, a low complication rate, and a low risk of relapse [29].

There are certain prognostic factors to be cognizant of when considering surgical management including estimated life expectancy, concomitant nonskeletal metastases, first metastasis to viscera, and use of systemic therapy. Patients with an estimated survival of at least 3–6 months or more should be considered surgical candidates [30].

Several methods may be utilized to determine risk of pathologic fracture and need for prophylactic surgery in metastatic bone disease. Among these are Mirel's [27] and Harrington's [31] criteria. Overall, however, significant untreatable pain and over 50 % destruction of cortical bone seem to be essential in evaluation [Fig. 5.1]. It has been shown that patients with prophylactic fixation exhibit shorter operative times, quicker recovery, and decreased morbidity than patients receiving repair after pathologic fracture [32]. Therefore, early evaluation of skeletal risk may help patients minimize complications associated with metastatic bone disease [8]. Better prediction of impending fracture and early orthopedic intervention may reduce frequency of progression to the undesired pathologic fracture [8].

Pathological fractures or impending fractures of weight bearing long bones are commonly treated with an intramedullary device so as to protect as much of the long bone as possible [Figs. 5.2 and 5.3]. A long-stem cemented arthroplasty device is also used for the same reason [26]. Fractures of long bones may also be treated with plates but this device is infrequently used in weight bearing bones due to greater soft tissue dissection required for placement. Further, plates are a load bearing device versus the load sharing ability of an intramedullary nail. Therefore,

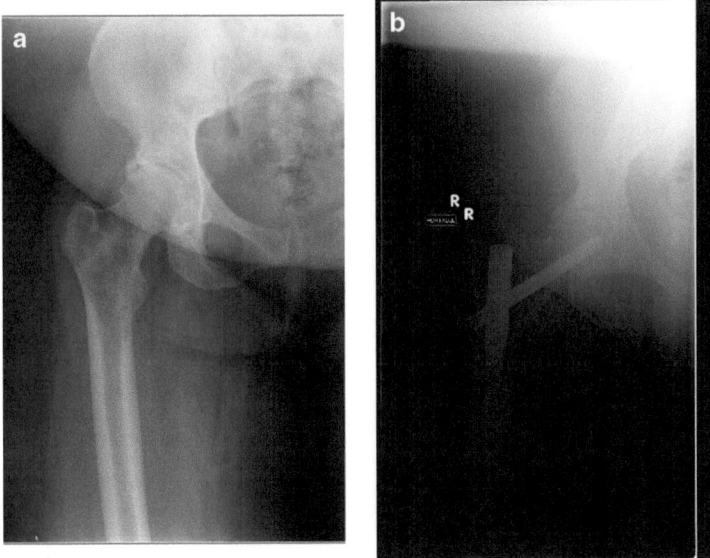

Fig. 5.1 (**a**) 58-year-old female with metastatic lesion involving more than 50 % of proximal femur. (**b**) Prophylactic nailing of femur performed for impending fracture

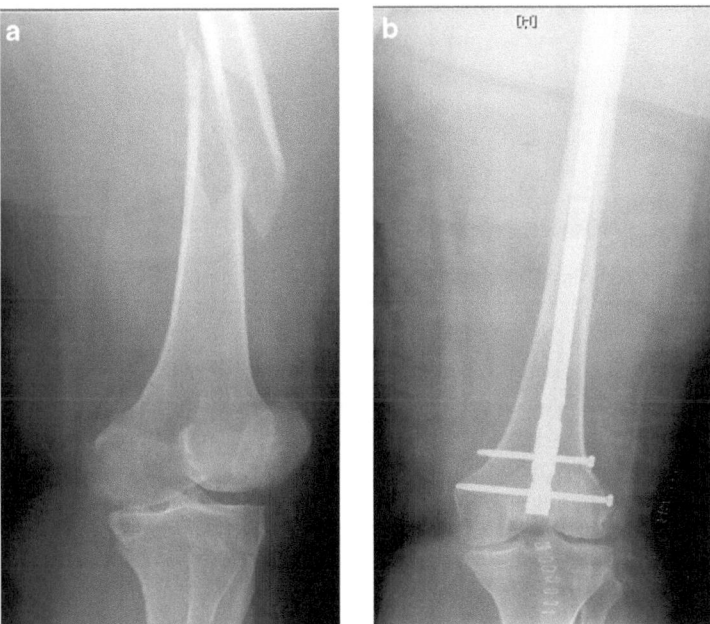

Fig. 5.2 (**a**) 56-year-old female with pathological fracture femur from breast metastasis (**b**) Following intramedullary nailing of femur

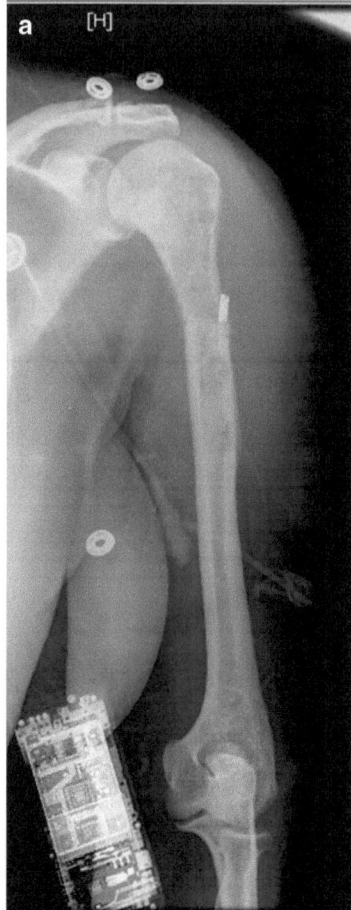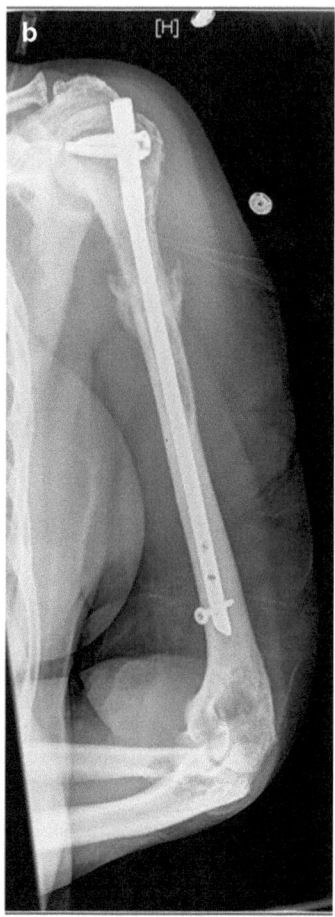

Fig. 5.3 (**a**) 59-year-old female with destructive lesion of humerus with minimally displaced pathological fracture (**b**) 6 months following intramedullary nailing of humerus

early weight bearing is not recommended with the use of a plate, delaying mobilization in an otherwise morbid population. Surgical fixation in patients with impending fracture is preventative for a complete break and does not provide curative therapy to the underlying cancer. The benefits of surgical fixation include pain relief, preservation or restoration of functional ability, and decreased overall morbidity. Risks of surgery include infection, bleeding, damage to surrounding neurovascular structures, and heterotopic ossification. There is also a major risk of embolic phenomena further increased due to the hypercoagulable state of malignancy. The area of fracture may subsequently be irradiated following fixation. This has been shown to reduce the incidence of hardware failure and need for re-operation [33]. Overall, however, bone metastases from breast cancer tend to respond better than other malignancies, with failure rates being below 10 % [7].

Vertebral compression fractures occur when metastatic deposits invade the vertebral column. Spinal cord compression is an extreme complication of metastatic cancer. It can lead to pain, spinal instability, neurological deficits, and a reduction in the patient's quality of life [34]. The cause of damage to the spinal cord from compression is multifactorial although two mechanisms predominate. These include direct compression causing a mass effect and a vascular injury resulting in irreversible cord ischemia [35]. Spinal cord compression causing neurologic deficits may be treated with chemotherapy, radiation, and surgery. There are reports of better outcomes following a combination of spinal decompression and radiation [35]. Although the surgical treatment has to be tailored to the needs of the patient, in principle it involves decompression of the neural tissue and partial vertebral resection along with stabilization using suitable implants and cages. In a select group of patients with pain and no neurological deficit kyphoplasty may be an option. Bone cement is utilized to fill the vertebral body following restoration of vertebral height with an inflatable balloon.

Morbidity

Bone is the most common site of breast cancer metastases and first distant relapse [36]. Around 70 % of patients who die with breast cancer have radiological evidence of bony involvement, whether or not this was known prior to death [1]. 29 % of those who develop bony metastases from breast cancer will experience one or more of the major complications, which include bone destruction, hypercalcemia, pathologic fracture, and spinal cord compression [8].

Bone destruction will lead to pathological fracture, hypercalcemia, or spinal cord compression. Pathologic fracture occurs in about 16 % of patients with bone metastases, and spinal cord compression occurs in about 3 % of these patients [8]. Although spinal cord compression is a relatively rare complication, it is one of the most morbid. Hypercalcemia is generally associated with poor outcomes, leading to a median of 3 months survival [8]. About 10 % of patients who have breast cancer will develop hypercalcemia, and 17 % of patients having breast cancer with bone metastases will develop hypercalcemia [8]. Additionally, hypercalcemia may be indicative of widespread bony involvement, as 85 % of metastatic breast cancer patients with hypercalcemia will have evidence of multiple bone metastases [8]. Pain due to bone metastases and orthopedic intervention is also an undesired result affecting quality of life and patient independence.

Morbidity due to surgery may also ensue in patients receiving operative treatment for metastatic bone disease. Some major risks include injury to the local neural and vascular structures, infection of the involved bone or surrounding soft tissue, and major blood loss. Another complication that is unique to this scenario includes seeding of more bone with the cancer cells. This may be a possible complication, especially in those receiving intramedullary instrumentation.

Concerns regarding breast cancer with bone metastases do not exclude financial burden both to the patient and the hospital. Skeletal-related events, which include any of the major complications associated with bone metastases, are associated with high costs and hospitalizations. Spinal cord compression poses the greatest financial burden, while pathologic fracture poses the least [37].

Survival

The overall survival of metastatic bone disease due to breast cancer has been determined to be dependent primarily on the location and the amount of metastases, whereas the patient age and type of orthopedic surgery performed shows minimal impact on survival [1]. The extent of metastatic disease and duration of associated symptoms have also been shown to be associated with survival [38]. Patients who exhibit solitary bone lesions have a 39 % 5-year survival rate [38]. In patients whose metastatic disease is confined to bone only, their median survival is 24 months. Median survival of those with first relapse of metastatic disease to bone is about 20 months [8]. This is significantly better than first relapse in the viscera, which has a median survival of 3 months [8]. If hypercalcemia ensues as a complication of the breast cancer or bony involvement, survival remains dismal at 3 months, and this usually indicates widespread disease [8]. Median survival following the diagnosis of spinal cord compression is 4 months, as surgical and radiotherapeutic intervention do not seem to affect survival in these patients [9]. Ambulation after diagnosis of spinal cord compression seems to be the only item to have effect on survival [9].

Currently there is no consensus in literature whether or not wide resection of metastatic bone tumor affects patient survival. The mean survival time after surgically fixing a pathologic fracture of the humerus is 8 months [29]. Survival at 1 year for patients with fixation of femoral fractures is 40 % [28].

Conclusions

Bone metastasis following breast cancer are not an uncommon occurrence and indicate disease progression. Patients have significant morbidity due to pain and impending or complete fractures. Bisphosphonates decrease osteoclastic activity and are administered to delay or prevent skeletal events. Surgery has a definite role in the treatment of pathological fractures and spinal cord compression. It helps improve quality of life but is not known to increase survival.

References

1. Wegener B, Schlemmer M, Stemmler J, Jansson V, Dürr HR, Pietschmann MF (2012) Analysis of orthopedic surgery of bone metastases in breast cancer patients. BMC Musculoskelet Disord 13:232. doi:10.1186/1471-2474-13-232

2. Quinn RH, Randall RL, Benevenia J, Berven SH, Raskin KA (2013) Contemporary management of metastatic bone disease: tips and tools of the trade for general practitioners. J Bone Joint Surg Am 95(20):1887–1895, Review

3. Edwards BK, Noone AM, Mariotto AB, Simard EP, Boscoe FP, Henley SJ, Jemal A, Cho H, Anderson RN, Kohler BA, Eheman CR, Ward EM (2013) Annual Report to the Nation on the status of cancer, 1975–2010, featuring prevalence of comorbidity and impact on survival among persons with lung, colorectal, breast, or prostate cancer. Cancer. doi:10.1002/cncr.28509, Epub ahead of print

4. Azim H, Azim HA Jr (2013) Targeting RANKL in breast cancer: bone metastasis and beyond. Expert Rev Anticancer Ther 13(2):195–201. doi:10.1586/era.12.177, Review

5. Greco C, Forte L, Erba P, Mariani G (2011) Bone metastases, general and clinical issues. Q J Nucl Med Mol Imaging 55(4):337–352

6. Ibrahim T, Farolfi A, Mercatali L, Ricci M, Amadori D (2013) Metastatic bone disease in the era of bone-targeted therapy: clinical impact. Tumori 99(1):1–9

7. Coleman RE (2006) Clinical features of metastatic bone disease and risk of skeletal morbidity. Clin Cancer Res 12:6243–6249

8. Coleman RE, Rubens RD (1987) The clinical course of bone metastases from breast cancer. Br J Cancer 55(1):61–66

9. Hill ME, Richards MA, Gregory WM, Smith P, Rubens RD (1993) Spinal cord compression in breast cancer: a review of 70 cases. Br J Cancer 68(5):969–973

10. Chambers AF, Groom AC, MacDonald IC (2002) Dissemination and growth of cancer cells in metastatic sites. Nat Rev Cancer 2(8):563–572, Review

11. Kang Y, Siegel PM, Shu W, Drobnjak M, Kakonen SM, Cordón-Cardo C, Guise TA, Massagué J (2003) A multigenic program mediating breast cancer metastasis to bone. Cancer Cell 3(6):537–549

12. Jezierska A, Motyl T (2009) Matrix metalloproteinase-2 involvement in breast cancer progression: a mini-review. Med Sci Monit 15(2):RA32–RA40, Review

13. Cabioglu N, Sahin AA, Morandi P, Meric-Bernstam F, Islam R, Lin HY, Bucana CD, Gonzalez-Angulo AM, Hortobagyi GN, Cristofanilli M (2009) Chemokine receptors in advanced breast cancer: differential expression in metastatic disease sites with diagnostic and therapeutic implications. Ann Oncol 20(6):1013–1019. doi:10.1093/annonc/mdn740, Epub 2009 Feb 23

14. Soria G, Ben-Baruch A (2008) The inflammatory chemokines CCL2 and CCL5 in breast cancer. Cancer Lett 267(2):271–285. doi:10.1016/j.canlet.2008.03.018, Epub 2008 Apr 24. Review

15. Harvey HA (1997) Issues concerning the role of chemotherapy and hormonal therapy of bone metastases from breast carcinoma. Cancer 80(8 Suppl):1646–1651, Review

16. Roodman GD (2004) Mechanisms of bone metastasis. N Engl J Med 350(16):1655–1664, Review

17. Clines GA, Guise TA (2008) Molecular mechanisms and treatment of bone metastasis. Expert Rev Mol Med 10:e7. doi:10.1017/S1462399408000616, Review

18. Guise TA, Mohammad KS, Clines G, Stebbins EG, Wong DH, Higgins LS, Vessella R, Corey E, Padalecki S, Suva L, Chirgwin JM (2006) Basic mechanisms responsible for osteolytic and osteoblastic bone metastases. Clin Cancer Res 12(20 Pt 2):6213s–6216s, Review

19. Elston CW, Blamey RW, Johnson J, Bishop HM, Haybittle JL, Griffiths K (1980) The relationship of oestradiol receptor (ER) and histological tumour differentiation with prognosis in human primary breast carcinoma. Eur J Cancer (Suppl 1):59–62

20. Lee YT (1985) Biochemical and hematological tests in patients with breast carcinoma: correlations with extent of disease, sites of relapse, and prognosis. J Surg Oncol 29(4):242–248

21. Roelofs AJ, Thompson K, Gordon S, Rogers MJ (2006) Molecular mechanisms of action of bisphosphonates: current status. Clin Cancer Res 12(20 Pt 2):6222s–6230s, Review

22. Wong M, Pavlakis N (2011) Optimal management of bone metastases in breast cancer patients. Breast Cancer (Dove Med Press) 3:35–60, eCollection 2011. Review

23. Pavlakis N, Schmidt R, Stockler M (2005) Bisphosphonates for breast cancer. Cochrane Database Syst Rev (3):CD003474. Review

24. Van Poznak CH, Temin S, Yee GC, Janjan NA, Barlow WE, Biermann JS, Bosserman LD, Geoghegan C, Hillner BE, Theriault RL, Zuckerman DS, Von Roenn JH, American Society of Clinical Oncology (2011) American Society of Clinical Oncology executive summary of the clinical practice guideline update on the role of bone-modifying agents in metastatic breast cancer. J Clin Oncol 29(9):1221–1227. doi:10.1200/JCO.2010.32.5209, Epub 2011 Feb 22. Erratum in: J Clin Oncol. 2011 Jun 1;29(16):2293

25. Hoskin PJ (2003) Bisphosphonates and radiation therapy for palliation of metastatic bone disease. Cancer Treat Rev 29(4):321–327, Review

26. Alvi HM, Damron TA (2013) Prophylactic stabilization for bone metastases, myeloma, or lymphoma: do we need to protect the entire bone? Clin Orthop Relat Res 471(3):706–714

27. Mirels H (1989) Metastatic disease in long bones. A proposed scoring system for diagnosing impending pathologic fractures. Clin Orthop Relat Res (249):256–264

28. Piccioli A, Rossi B, Scaramuzzo L, Spinelli MS, Yang Z, Maccauro G (2014) Intramedullary nailing for treatment of pathologic femoral fractures due to metastases. Injury 45(2):412–417. doi:10.1016/j.injury.2013.09.025, Epub 2013 Sep 19

29. Piccioli A, Maccauro G, Rossi B, Scaramuzzo L, Frenos F, Capanna R (2010) Surgical treatment of pathologic fractures of humerus. Injury 41(11):1112–1116. doi:10.1016/j.injury. 2010.08.015, Epub 2010 Sep 9

30. Weber KL, Lewis VO, Randall RL, Lee AK, Springfield D (2004) An approach to the management of the patient with metastatic bone disease. Instr Course Lect 53:663–676, Review

31. Harrington KD (1986) Impending pathologic fractures from metastatic malignancy: evaluation and management. Instr Course Lect 35:357–381

32. Ward WG, Spang J, Howe D, Gordan S (2000) Femoral recon nails for metastatic disease: indications, technique, and results. Am J Orthop (Belle Mead NJ) 29(9 Suppl):34–42

33. Damron TA, Sim FH (2000) Surgical treatment for metastatic disease of the pelvis and the proximal end of the femur. Instr Course Lect 49:461–470, Review

34. Klimo P Jr, Kestle JR, Schmidt MH (2003) Treatment of metastatic spinal epidural disease: a review of the literature. Neurosurg Focus 15(5):E1, Review

35. Patchell RA, Tibbs PA, Regine WF, Payne R, Saris S, Kryscio RJ, Mohiuddin M, Young B (2005) Direct decompressive surgical resection in the treatment of spinal cord compression caused by metastatic cancer: a randomised trial. Lancet 366(9486):643–648

36. McNeil BJ (1984) Value of bone scanning in neoplastic disease. Semin Nucl Med 14 (4):277–286, Review

37. Hagiwara M, Delea TE, Chung K (2014) Healthcare costs associated with skeletal-related events in breast cancer patients with bone metastases. J Med Econ 17(3):223–230. doi:10. 3111/13696998.2014.890937

38. Dürr HR, Müller PE, Lenz T, Baur A, Jansson V, Refior HJ (2002) Surgical treatment of bone metastases in patients with breast cancer. Clin Orthop Relat Res 396:191–196

Chapter 6
Molecular Targeted Therapy for Brain Metastatic Breast Cancers: Current Updates

Aamir Ahmad and Fazlul H. Sarkar

Abstract Metastatic breast cancers are difficult to manage, and brain metastatic breast cancers are particularly lethal with poor overall survival and outcomes. Our understanding of factors that facilitate brain metastasis of breast cancers is very poor. However, a number of studies in the last few years have focused on brain metastatic breast cancers; a number of therapeutic targets have been tested, and few therapeutic regimes have also been evaluated in phase I/II settings. We discuss here some of the most recent developments in the field. We will limit our discussion to the reports from past one year only to make this article most up-to-date with no repetition of information that has already been reviewed in the available literature.

Keywords Breast cancer • Brain metastasis • Blood–brain barrier • Estrogen receptor-negative breast cancer • HER2-positive breast cancer

Introduction

Breast cancer metastasizes to the brain in approximately 10–15 % of patients and is usually associated with very poor prognosis and survival [1]. New data suggests that the incidence of breast cancer brain metastasis might be even higher—15 % to 40 % [2, 3]. Brain metastases, in general, are the most frequent brain tumor type, and the breast happens to be one of the most common primary origins of tumors that eventually metastasize in the brain [4], and thus, brain metastasis of breast cancer is highly lethal. There are not many treatment options because most chemotherapies are not useful which is in part due to the presence of the blood–brain barrier (BBB)

A. Ahmad
Department of Pathology, Karmanos Cancer Institute, Wayne State University School of Medicine, Detroit, MI, USA

F.H. Sarkar (✉)
Department of Pathology, Karmanos Cancer Institute, Wayne State University School of Medicine, Detroit, MI, USA

Department of Oncology, Karmanos Cancer Institute, Wayne State University School of Medicine, 740 Hudson Webber Cancer Research Center, 4100 John R Street, Detroit, MI 48201, USA
e-mail: fsarkar@med.wayne.edu

© Springer International Publishing Switzerland 2014
S. Sethi, *miRNAs and Target Genes in Breast Cancer Metastasis*,
SpringerBriefs in Cancer Research, DOI 10.1007/978-3-319-08162-5_6

where most of the drugs are unable to cross this barrier. Whole brain radiotherapy is often the first line of defense, and there is evidence to suggest the efficacy of this approach to certain extent [5].

The mechanism of brain metastasis of breast cancer is not very clearly understood, and a number of studies have looked at the possible factors that might help us gain a better understanding of the underlying principles that lead to migration of breast tumor cells into the brain [2, 6–9]. A number of review articles have been written on this topic [1, 10–20]. In this article, we will limit our discussion to the most recent advances in the field with a focus on studies reported in last one year only.

Receptor Expression and Breast Cancer Brain Metastasis

Breast cancer subtypes are frequently classified based on the presence or absence of receptors such as estrogen receptor (ER), progesterone receptor (PR), and epidermal growth factor receptor 2 (HER2/ErbB2). In a study [21] that tried to establish a predictability of breast cancer brain metastasis based on the expression of ER/PR/HER2, no clear connection was established. Of the 6 patient samples studied by immunohistochemistry, four were ER-positive, while five each were positive for PR and HER2. This suggests that one or more of these three classical receptors are expressed in the patients' samples. When summarized for all the receptors at the same time, it was reported that four samples were positive for all the three receptors; one was positive for PR and HER2 but negative for ER, while one was positive for PR but negative for ER and HER2. These observations suggest a clear heterogeneity within the population of breast cancer patients with brain metastasis. However, these are some very preliminary results with just 6 samples, and, clearly, bigger sample size studies are warranted which may provide better insights as to the role of ER/PR/HER2 in the brain cancer metastasis of breast cancer.

From the perspective of better clinical management, it would be highly desirable if the development of brain metastases could somehow be predicted. With this goal in mind, Rudat et al. [22] looked for factors, in particular, the expression of ER/PR/HER2, to see if the presence of any of these in local or locoregional disease could be a predictor for future brain metastasis. To accomplish this, medical records of 352 breast cancer patients were analyzed retrospectively. It was reported that ER-negative, PR-negative, and triple negative status were the highly significant risk factors for developing brain metastasis. In a subgroup of 168 patients where follow-up data was available for at least 24 months, 49 patients were reported to exhibit brain metastasis as the initial metastatic event. This means that more than 29 % of breast cancer patients developed brain metastasis. The major conclusion of this study was that ER-negative status exhibits an almost 50 % risk of developing brain metastasis in breast cancer patients.

Since HER2-positive breast tumors are also believed to metastasize to the brain, a study was designed to assess gene expression signature differences between

HER2-positive brain metastatic tumors and the HER2-positive nonmetastatic primary breast tumors [23]. Not all HER2-positive breast cancers exhibited brain metastases, and, therefore, such analysis might help identify target genes that are overexpressed or lost which may lead to the development of brain metastases. This exercise led to the identification of BRCA1 as the factor that is lost in HER2-positive tumors that metastasize to the brain. Thus, a BRCA1 deficient-like (BD-L) signature appears to correlate with breast cancer brain metastasis in HER2-positive breast cancers. Interestingly, this correlation is higher in HER2-positive/ER-negative primary tumors, as compared to either HER2-positive/ER-positive of HER2-negative/ER-positive primary breast tumors. This study reveals loss of BRCA1 in HER2 expressing breast tumors as a predictor of brain metastasis. There also seems to be some link between the losses of ER expression similar to the study discussed above.

In yet another recent study [24] focused on risk factor, particularly the subtypes and the age, it was determined that brain metastases of breast cancer vary between different subtypes among different age groups. For example, it was found that the risk of brain metastasis in patients aged 35 or younger is independent of breast cancer subtype. However, in patients over 36 years of age, HER2 overexpressing and the triple negative breast cancer subtypes pose a higher risk of brain metastasis. In an earlier study on the subject [25], analysis of data from 219 patients suggested that the onset of brain metastases correlated with younger age and 43 % of patients under 40 years of age presented with brain metastasis. The ER-negative status was also found to be correlated with brain metastasis as 38 % ER-negative patients developed brain metastasis, compared to only 14 % ER-positive ones. Furthermore, combining these two factors revealed that patients under 50 years of age and with ER-negative phenotype had 53 % chances of presenting with brain metastases and that the younger breast cancer patients are at higher risk of developing brain metastasis which is also supported by an early analysis done more than three decades ago [26]. This study suggested that the median age of breast cancer patients presenting metastasis to central nervous system is 5 years less than the patients without [24]. Evidence of brain metastasis at the initial diagnosis of breast cancer is rare. In such a unique group of patients where initial diagnosis of breast cancer is simultaneous with brain metastasis, longer survival has been reported which was also correlated with younger age [27].

Therapy

Not many therapeutic options are available once breast cancer metastasizes to the brain. Whole brain irradiation and surgical interventions are commonly practiced [28], but a relapse is very common. Whole brain irradiation simultaneously with targeted intervention such as the use of lapatinib has also been tested in phase I setting [29], but the results are not very conclusive. Lapatinib is a dual inhibitor: it targets both EGFR and HER2, and there is evidence for its efficacy against brain metastatic breast cancers [30, 31]. Many different drugs have been proposed and/or evaluated against brain metastatic breast cancer, and we will discuss some of the most recent reports here in this section.

Mutlu and Buyukcelik tested a weekly combined administration of trastuzumab and vinorelbine in three breast cancer patients with brain metastasis [32]. The progression-free survival was reported to be 12, 16, and 9 months for the three patients. The three studied patients were similar in a way that they all had overexpression of HER2. However, patient 1 was positive for ER and PR; patient 2 was ER-negative but PR-positive; and patient 3 was negative for both ER and PR. Evidently, the three patients had heterogeneity with respect to receptor expression, and these results have significant limitation. Further, patient 1 had liver and bone metastases before brain metastasis while patient 2 had liver metastasis prior to brain metastasis. These patients were administered trastuzumab and vinorelbine for different time periods because of multiple reasons and also were treated with other drugs prior to this treatment. Patient 1 did not show any signs of brain metastasis after treatment; however, she was taken off from the regimen because of progression of bone metastasis. Patient 2 had to be taken off because of progression of liver as well as the brain metastases, while patient 3 succumbed to cerebrovascular embolic event. These cases, thus, were very different from each other and, accordingly, responded differently to the combination treatment as well. A more robust study with larger number of patients might be necessary before making a case for the efficacy of this treatment procedure combining trastuzumab and vinorelbine.

The research group led by Patricia Steeg tested the effect of temozolomide on the experimental brain metastasis of breast cancer [33]. Temozolomide was chosen because this drug is known to be brain permeable and is used for the treatment of primary brain tumors. This drug, however, acts in a O^6-methylguanine-DNA methyltransferase (MGMT)-dependent manner, where it is effective only when there is low to no MGMT activity. Using tissue microarrays, it was determined that a majority of brain metastases, 58 %, had low MGMT expression. This supported the rationale of using temozolomide in the experimental model. When MGMT-negative breast cancer cells were used to experimentally mimic brain metastases, the use of temozolomide was found to be very efficient. Doses of 5, 10, 25, and 50 mg/kg temozolomide were found to completely block the appearance of brain metastases when treatment was initiated 3 days postinoculation of cells, and the drug was administered 5 days a week. A lower dose, 1 mg/kg, was found to be effective as well, but the inhibition of large brain metastases was only 68 %; however, a lower dose of 0.5 mg/kg was totally ineffective. As expected, temozolomide treatment was ineffective when MGMT-expressing breast cancer cells were used in this experimental brain metastasis model. Also, a delayed treatment, such as 18 or 24 days postinjection of cells, was found to be ineffective which means that the "homing" of metastatic cells might be a point of no return. This, in turn, means that we need to evaluate risk factors for brain metastasis so that an effective therapy can be delivered in patients with higher risk or propensity for the development of brain metastasis, and the treatment strategy must be implemented much before the cells actually cross the BBB.

Whole brain irradiation is a preferred approach for treating brain metastatic breast cancers. As discussed above, the two subtypes of breast cancer that have high propensity to metastasize to the brain are HER2-positive and the triple negative breast cancers. Wu et al. [34] compared the efficacy of whole brain irradiation in triple negative breast cancers vs. non-triple negative breast cancers. While the efficiency of irradiation was similar against both subtypes, triple negative breast cancers were associated with poor survival rates: 6.9 months for triple negative vs. 17 months for non-triple negative breast cancers. It appears that the triple negative breast cancers with brain metastasis are much more aggressive than those that are positive for one of the receptors. Triple negative breast cancers that migrate to the brain are aggressive, partly because of the absence of any validated therapeutic target; thus, it might be assumed that the HER2-overexpressing breast cancers might be easier to manage because of the available targeted agents. However, HER2-overexpressing breast cancers have their own unique challenges. First of all, there is the phenomenon of acquired drug resistance [35] where the tumors first respond to the targeted treatment but become resistant to the very same treatment with continued administration. Also, there is evidence that when the primary HER2-overexpressing breast tumors are treated with HER2-targeting agent trastuzumab, there is an increased risk of developing brain metastases [36]. In this retrospective study of 132 patients, it was observed that 43.3 % HER2-positive patients developed brain metastases when initially treated with trastuzumab. On the other hand, only 26.2 % HER2-positive patients from the control group developed brain metastases. The control group was not treated with trastuzumab initially. This represents a complex scenario where targeted therapy might put patients at increased risk of developing brain metastasis in the future.

As evident from the discussion in this section, a number of options have been considered for the management of breast cancer brain metastases, but no definite treatment has yet been established. In addition to those discussed above, a few other therapeutic targets/drugs have also been recently tested [37–41]. A case report has recently provided evidence of acute encephalopathy in a patient with breast cancer brain metastasis when treated with pegylated liposomal doxorubicin [42]. The reason for such condition was postulated to be the partial disruption of BBB. This reminds us of the immense challenges associated with treating brain metastases and the inadvertent side effects of the chosen treatment. Great caution needs to be exercised while treating breast cancer brain metastasis with utmost consideration of maintaining the normal brain function and physiology.

Novel Targets for Targeting Brain Metastatic Breast Cancers

Genomic and Epigenomic Targets

With the objective of identifying genomic and epigenomic events that are correlated with brain metastasis of breast cancer, deep genomic analysis was performed [2] where breast cancer brain metastasis samples were compared to nonneoplastic breast and brain specimens. As part of genomic events, frequent large chromosomal gains in 1q, 5p, 8q, 11q, and 20q and frequent deletions in 8p, 17p, 21p, and Xq were observed. Genes ATAD2, BRAF, DERL1, DNMTRB, and NEK2A were found to be overexpressed, while ATM, CRYAB, and HSPB2 were deleted or silenced. An involvement of cell cycle modulators was evident in brain metastases, and AURKA, AURKB, and FOXM1 were overexpressed. Among the various subtypes, the most common subtypes with brain metastases were identified to be luminal B, HER2-positive/ER-negative, and basal-like breast cancers. As for the epigenetic changes, hypermethylation and downregulation of PENK, EDN3, and ITGAM were evident in brain metastases. These results are suggestive of multiple physiologic events that aid in the brain metastasis of breast cancers. In an earlier study [43] on gene expression profiling of brain metastatic triple negative breast cancers, periplakin and mitogen-activated protein kinase 13 were determined to be associated with brain metastatic profile. In vitro, they were found to influence cell growth and motility. It would be interesting to validate their role in brain metastasis in vivo.

Serpins

A recent work in the laboratory of Joan Massague [44] has described an interaction between the brain stroma-derived plasmin and the cancer cell-derived serpins. The reactive brain stroma attempts to block the invading cancer cells through release of plasmin. This is countered by the release of serpins by the invading cancer cells which function as a shield for these metastatic cells providing them protection against Fas-dependent death. Brain metastatic breast cancer cells undergo Fas-dependent death in case they are not protected by the serpins. Mechanistically, serpins protect cancer cells by blocking plasmin-mediated degradation of L1CAM. It is interesting to note that while plasmins themselves have been implicated in the proliferation and invasion of cancer cells through their degradation of extracellular matrix, it is the levels of serpins that have been correlated with poor outcome in human cancer patients [44, 45]. It is not surprising that the brain metastatic lung and breast cancer cells were observed to express high levels of plasminogen activator inhibitory serpins, such as neuroserpin and serpin B2. It, thus, appears that selective targeting of serpins in primary breast tumors might inhibit their ability to survive in the hostile brain microenvironment even if they metastasize.

αB-Crystallin

Malin et al. [46] evaluated the role of αB-crystallin in brain metastasis of triple negative breast cancers. Since αB-crystallin is predominantly expressed in triple negative breast cancers and such cancers also present a higher risk of brain metastasis, an immunohistochemical analysis was first done which demonstrated a correlation between αB-crystallin expression and poor survival after brain metastasis. Overexpression of αB-crystallin aided the migration of triple negative breast cancer cells across BBB in an experimental model and promoted brain metastasis. Silencing of αB-crystallin inhibited brain metastases. In vivo studies for brain metastasis were performed in an orthotopic model where triple negative breast cancer cells were injected into mammary glands of mice. Metastases were observed in different organs and were not just confined to the brain. This study highlights a role of αB-crystallin in the overall metastatic potential of triple negative breast cancers and may have special implications in brain metastasis.

Angiopoietin-2

Angiopoietin-2 has been suggested as a therapeutic target [47] based on its ability to assist in the brain metastasis of triple negative breast cancer cells. Angiopoietin-2 seems to play an important role in the disruption of BBB which is an essential step in the brain metastatic cascade. This was confirmed by the observation that secreted angiopoietin-2 resulted in enhanced BBB permeability, whereas the use of neutralizing angiopoietin-2 agent, trebananib, afforded protection to BBB and resulted in the inhibition of brain metastasis of breast cancer cells.

ADAM8

ADAM8, the "a disintegrin and metalloproteinase domain-containing protein 8," is a transmembrane protein that is involved in cell adhesion and migration. Its expression is particularly high in primary breast tumors, compared to adjacent normal tissue, and its levels are higher in the serum of breast cancer patients [48]. Its expression levels are further increased in triple negative breast cancer cells when cultured in three-dimensional cultures. Knockdown of ADAM8 was found to be associated with decreased invasive potential and colony formation [48]. In relation to brain metastasis, it was observed that the metastasis to the brain was significantly reduced in mice that were injected with triple negative breast cancer cells MDA-MB-231 knocked down for ADAM8. Whereas all control mice (6/6) had brain metastasis, only one mouse in the ADAM8 knockdown group had brain metastasis. Such knockdown of ADAM8 was also found to be correlated with reduced circulating tumor cells (CTCs) and the overall tumor burden. CTCs are now believed to be related to the outcome in HER2-positive brain metastatic breast cancers [49]. In summary, the in vitro as well as in vivo evidence presented in this study [48] makes a case for further investigation of ADAM8 regarding its role in brain metastasis of breast cancer cells.

Tumor Microenvironment

One of the least understood process is how breast cancer cells cross BBB in order to reach the brain. BBB is a specialized vascular interface that restricts the transport of most compounds into the brain [50]. Consequently, crossing BBB for effective delivery of putative therapeutic drugs is also an enormous challenge [51]. Choi and co-workers [52] looked at the role played by cancer-associated fibroblasts (CAFs) in the process of brain metastasis of breast cancer cells. In the recent years, it has been realized that the stroma, surrounding the tumor, cooperates with the tumor cells and, in particular, the CAFs from the stoma help in the aggressive display by tumor cells facilitating the processes of angiogenesis, invasion, and metastasis. Tumor microenvironment is important for the metastasis of breast cancer cells [53]. Using three-dimensional cultures, Choi et al. were able to demonstrate the potentiation of invasive potential of breast cancer cells by CAFs [52]. The invasive potential was determined to be increased 1.78- and 1.83-folds after 6 and 9 days of coculture, respectively. An in vitro BBB model was developed using human brain microvascular endothelial cells (HBMECs). While control MDA-MB-231 breast cells could not disrupt BBB, coculture of these cells with CAFs leads to BBB disruption which was similar to cells derived from MDA-MB-231 cells that preferentially metastasize in the brain. These results are indicative of a role of CAFs in the crossing of BBB by breast cancer cells. Such information will be important when designing novel therapies for the prevention and/or treatment of breast cancer brain metastases.

Wang et al. [54] provided evidence for the role of astrocytes in determining the metastatic potential of breast cancer cells. They observed an increased metastasis of breast cancer cells to the brain when the cells were preconditioned with astrocytes media. Mechanistically, a role of MMP-2 and MMP-9 in such an increased metastasis potential was proposed because the blockage of these factors in conditioned media attenuated the metastatic potential. In conclusion, a number of interesting potential targets have been proposed in the last one year that might play a role in the brain metastasis of breast cancers. Additional targets such as type I insulin-like growth factor receptor [55] have also been evaluated which are not discussed here in detail because they are just outside the time frame of studies discussed here. All of these promising leads need to be evaluated further in preclinical and clinical settings.

Conclusions and Perspectives

Brain metastatic breast cancers are difficult to treat largely because of the lack of knowledge on mechanism(s) of their onset and progression. BBB represents a major roadblock in the way of designing effective therapeutic regimes. In order to be effective against brain metastatic breast cancer, any putative drug must cross the BBB. However, passing BBB will just be the first hurdle, as suggested by the failed phase II trial [56] of a drug patupilone that can cross BBB but has no significant antitumor activity. Brain metastatic breast cancers are, therefore, complex to understand. They are further complicated by the new evidence that once metastasized in the brain, breast cancer cells exhibit neural characteristics [57], i.e., they start expressing factors that are known to be markers of neural/brain cells. Perhaps, this is one way by which breast cancers "adapt" to their new microenvironment. A survey of literature reveals a spurt in studies on the brain metastasis of breast cancer in the last couple of years. This is promising but the momentum needs to be maintained, and many more clinical studies need to be planned so that there is some hope for patients suffering from this deadly disease, in order to improve the overall survival.

References

1. Weil RJ, Palmieri DC, Bronder JL et al (2005) Breast cancer metastasis to the central nervous system. Am J Pathol 167:913–920
2. Salhia B, Kiefer J, Ross JT et al (2014) Integrated genomic and epigenomic analysis of breast cancer brain metastasis. PLoS One 9:e85448
3. Tosoni A, Franceschi E, Brandes AA (2008) Chemotherapy in breast cancer patients with brain metastases: have new chemotherapic agents changed the clinical outcome? Crit Rev Oncol Hematol 68:212–221
4. Gallego Perez-Larraya J, Hildebrand J (2014) Brain metastases. Handb Clin Neurol 121:1143–1157
5. Yoshioka S, Hojo S, Toyoda Y et al (2013) The efficacy of early diagnosis of brain metastasis and systemic treatment after radiotherapy in patients with metastatic breast cancer. Gan To Kagaku Ryoho 40:2381–2383
6. Engin HB, Guney E, Keskin O et al (2013) Integrating structure to protein-protein interaction networks that drive metastasis to brain and lung in breast cancer. PLoS One 8:e81035
7. Gojis O, Kubecova M, Rosina J et al (2013) Expression of selected proteins in breast cancer brain metastases. Folia Histochem Cytobiol 51:213–218
8. Gilkes DM, Semenza GL (2013) Role of hypoxia-inducible factors in breast cancer metastasis. Future Oncol 9:1623–1636
9. Zhao H, Jin G, Cui K et al (2013) Novel modeling of cancer cell signaling pathways enables systematic drug repositioning for distinct breast cancer metastases. Cancer Res 73:6149–6163
10. Palmieri D, Smith QR, Lockman PR et al (2006) Brain metastases of breast cancer. Breast Dis 26:139–147
11. Palmieri D, Chambers AF, Felding-Habermann B et al (2007) The biology of metastasis to a sanctuary site. Clin Cancer Res 13:1656–1662
12. Gril B, Evans L, Palmieri D et al (2010) Translational research in brain metastasis is identifying molecular pathways that may lead to the development of new therapeutic strategies. Eur J Cancer 46:1204–1210
13. Steeg PS, Camphausen KA, Smith QR (2011) Brain metastases as preventive and therapeutic targets. Nat Rev Cancer 11:352–363
14. Sleeman J, Steeg PS (2010) Cancer metastasis as a therapeutic target. Eur J Cancer 46:1177–1180
15. Davies MA (2012) Targeted therapy for brain metastases. Adv Pharmacol 65:109–142

16. Soffietti R, Trevisan E, Ruda R (2012) Targeted therapy in brain metastasis. Curr Opin Oncol 24:679–686

17. Mehta AI, Brufsky AM, Sampson JH (2013) Therapeutic approaches for HER2-positive brain metastases: circumventing the blood-brain barrier. Cancer Treat Rev 39:261–269

18. Gil-Gil MJ, Martinez-Garcia M, Sierra A et al (2014) Breast cancer brain metastases: a review of the literature and a current multidisciplinary management guideline. Clin Transl Oncol 16:436–446

19. Dawood S, Gonzalez-Angulo AM (2013) Progress in the biological understanding and management of breast cancer-associated central nervous system metastases. Oncologist 18:675–684

20. Kuo AH, Clarke MF (2013) Identifying the metastatic seeds of breast cancer. Nat Biotechnol 31:504–505

21. Rao PS, Labhart M, Mayhew SL et al (2014) Heterogeneity in the expression of receptors in the human breast cancer metastasized to the brain. Tumour Biol. doi:10.1007/s13277-014-1979-9

22. Rudat V, El-Sweilmeen H, Brune-Erber I et al (2014) Identification of breast cancer patients with a high risk of developing brain metastases: a single-institutional retrospective analysis. BMC Cancer 14:289

23. McMullin RP, Wittner BS, Yang C et al (2014) A BRCA1 deficient-like signature is enriched in breast cancer brain metastases and predicts DNA damage-induced poly (ADP-ribose) polymerase inhibitor sensitivity. Breast Cancer Res 16:R25

24. Hung MH, Liu CY, Shiau CY et al (2014) Effect of age and biological subtype on the risk and timing of brain metastasis in breast cancer patients. PLoS One 9:e89389

25. Evans AJ, James JJ, Cornford EJ et al (2004) Brain metastases from breast cancer: identification of a high-risk group. Clin Oncol (R Coll Radiol) 16:345–349

26. Tsukada Y, Fouad A, Pickren JW et al (1983) Central nervous system metastasis from breast carcinoma. Autopsy study. Cancer 52:2349–2354

27. Wiksyk B, Nguyen DH, Alexander C et al (2014) Population-based analysis of treatment and survival in women presenting with brain metastasis at initial breast cancer diagnosis. Am J Clin Oncol. doi:10.1097/COC.0000000000000055

28. Jo KI, Im YH, Kong DS et al (2013) Gamma knife radiosurgery for brain metastases from breast cancer. J Korean Neurosurg Soc 54:399–404

29. Lin NU, Freedman RA, Ramakrishna N et al (2013) A phase I study of lapatinib with whole brain radiotherapy in patients with Human Epidermal Growth Factor Receptor 2 (HER2)-positive breast cancer brain metastases. Breast Cancer Res Treat 142:405–414

30. Cameron D, Casey M, Press M et al (2008) A phase III randomized comparison of lapatinib plus capecitabine versus capecitabine alone in women with advanced breast cancer that has progressed on trastuzumab: updated efficacy and biomarker analyses. Breast Cancer Res Treat 112:533–543

31. Oktay E, Yersal O, Meydan N et al (2013) Nearly complete response of brain metastases from HER2 overexpressing breast cancer with lapatinib and capecitabine after whole brain irradiation. Case Rep Oncol Med 2013:234391

32. Mutlu H, Buyukcelik A (2014) The combination of weekly trastuzumab plus vinorelbine may be preferable regimen in HER-2 positive breast cancer patients with brain metastasis. J Oncol Pharm Pract. doi:10.1177/1078155214531514

33. Palmieri D, Duchnowska R, Woditschka S et al (2014) Profound prevention of experimental brain metastases of breast cancer by temozolomide in an MGMT-dependent manner. Clin Cancer Res 20(10):2727–2739

34. Wu X, Luo B, Wei S et al (2013) Efficiency and prognosis of whole brain irradiation combined with precise radiotherapy on triple-negative breast cancer. J Cancer Res Ther 9(Suppl):S169–S172

35. Ahmad A, Sarkar FH (2013) Current understanding of drug resistance mechanisms and therapeutic targets in HER2 overexpressing breast cancers. In: Ahmad A (ed) Breast cancer metastasis and drug resistance, vol 1. Springer, New York, pp 261–274

36. Kaplan MA, Ertugrul H, Firat U et al (2014) Brain metastases in HER2-positive metastatic breast cancer patients who received chemotherapy with or without trastuzumab. Breast Cancer. doi:10.1007/s12282-013-0513-z
37. Zhang S, Huang WC, Zhang L et al (2013) SRC family kinases as novel therapeutic targets to treat breast cancer brain metastases. Cancer Res 73:5764–5774
38. Nakayama A, Takagi S, Yusa T et al (2013) Antitumor activity of TAK-285, an investigational, non-Pgp SUBSTRATE HER2/EGFR kinase inhibitor, in cultured tumor cells, mouse and rat xenograft tumors, and in an HER2-positive brain metastasis model. J Cancer 4:557–565
39. Mittapalli RK, Liu X, Adkins CE et al (2013) Paclitaxel-hyaluronic nanoconjugates prolong overall survival in a preclinical brain metastases of breast cancer model. Mol Cancer Ther 12:2389–2399
40. Gupta P, Adkins C, Lockman P et al (2013) Metastasis of breast tumor cells to brain is suppressed by phenethyl isothiocyanate in a novel metastasis model. PLoS One 8:e67278
41. Boothe D, Young R, Yamada Y et al (2013) Bevacizumab as a treatment for radiation necrosis of brain metastases post stereotactic radiosurgery. Neuro Oncol 15:1257–1263
42. Baker M, Markman M, Niu J (2014) Pegylated liposomal doxorubicin-induced acute transient encephalopathy in a patient with breast cancer: a case report. Case Rep Oncol 7:228–232
43. Choi YK, Woo SM, Cho SG et al (2013) Brain-metastatic triple-negative breast cancer cells regain growth ability by altering gene expression patterns. Cancer Genomics Proteomics 10:265–275
44. Valiente M, Obenauf AC, Jin X et al (2014) Serpins promote cancer cell survival and vascular co-option in brain metastasis. Cell 156:1002–1016
45. Berger DH (2002) Plasmin/plasminogen system in colorectal cancer. World J Surg 26:767–771
46. Malin D, Strekalova E, Petrovic V et al (2014) alphaB-crystallin: a novel regulator of breast cancer metastasis to the brain. Clin Cancer Res 20:56–67
47. Avraham HK, Jiang S, Fu Y et al (2014) Angiopoietin-2 mediates blood-brain barrier impairment and colonization of triple-negative breast cancer cells in brain. J Pathol 232:369–381
48. Romagnoli M, Mineva ND, Polmear M et al (2014) ADAM8 expression in invasive breast cancer promotes tumor dissemination and metastasis. EMBO Mol Med 6:278–294
49. Pierga JY, Bidard FC, Cropet C et al (2013) Circulating tumor cells and brain metastasis outcome in patients with HER2-positive breast cancer: the LANDSCAPE trial. Ann Oncol 24:2999–3004
50. Adkins CE, Mittapalli RK, Manda VK et al (2013) P-glycoprotein mediated efflux limits substrate and drug uptake in a preclinical brain metastases of breast cancer model. Front Pharmacol 4:136
51. Connell JJ, Chatain G, Cornelissen B et al (2013) Selective permeabilization of the blood-brain barrier at sites of metastasis. J Natl Cancer Inst 105:1634–1643
52. Choi YP, Lee JH, Gao MQ et al (2014) Cancer-associated fibroblast promote transmigration through endothelial brain cells in three-dimensional in vitro models. Int J Cancer. doi:10.1002/ijc.28848
53. Karnoub AE, Dash AB, Vo AP et al (2007) Mesenchymal stem cells within tumour stroma promote breast cancer metastasis. Nature 449:557–563
54. Wang L, Cossette SM, Rarick KR et al (2013) Astrocytes directly influence tumor cell invasion and metastasis in vivo. PLoS One 8:e80933
55. Saldana SM, Lee HH, Lowery FJ et al (2013) Inhibition of type I insulin-like growth factor receptor signaling attenuates the development of breast cancer brain metastasis. PLoS One 8:e73406
56. Peereboom DM, Murphy C, Ahluwalia MS et al (2014) Phase II trial of patupilone in patients with brain metastases from breast cancer. Neuro Oncol 16:579–583
57. Neman J, Termini J, Wilczynski S et al (2014) Human breast cancer metastases to the brain display GABAergic properties in the neural niche. Proc Natl Acad Sci USA 111:984–989

Index

© Springer International Publishing Switzerland 2014
S. Sethi, *miRNAs and Target Genes in Breast Cancer Metastasis*,
SpringerBriefs in Cancer Research, DOI 10.1007/978-3-319-08162-5